THE MYTH OF THE NATURAL LABORATORY

THE MYTH OF THE NATURAL LABORATORY

Science, Empire, and Their Derangements on Pitcairn and Norfolk Islands

Adrian Young

University of Hawai'i Press
HONOLULU

Printed in the United States of America

First printed, 2026

Library of Congress Cataloging-in-Publication Data

Names: Young, Adrian (Historian), author.
Title: The myth of the natural laboratory : science, empire, and their derangements on Pitcairn and Norfolk islands / Adrian Young.
Other titles: Science, empire, and their derangements on Pitcairn and Norfolk islands
Description: Honolulu : University of Hawai'i Press, [2026] | Includes bibliographical references and index.
Identifiers: LCCN 2025020250 | ISBN 9798880700721 (hardback) | ISBN 9798880700738 (trade paperback) | ISBN 9798880700752 (epub) | ISBN 9798880700769 (kindle edition) | ISBN 9798880700745 (pdf)
Subjects: LCSH: Ethnology—Pitcairn Islands—History. | Ethnology—Norfolk Island—History. | Pacific Islanders—Study and teaching. | Imperialism and science.
Classification: LCC DU800 .Y693 2026
LC record available at https://lccn.loc.gov/2025020250

Cover art: *Landing in Bounty Bay*, c. 1830. Oil painting. Attributed to Richard Brydges Beechey, based on a drawing by Frederick William Beechey. Mitchell Library, State Library of New South Wales, ML 114.

University of Hawai'i Press books are printed on acid-free paper and meet the guidelines for permanence and durability of the Council on Library Resources.

Contents

Acknowledgments

The space of a few pages is hardly enough to acknowledge all the debts I have accrued after more than a decade of writing and research. My work has been shaped by hundreds of people and thousands of interactions, big and small. But it is perhaps fitting that this book, a history of research on and about Pitcairn and Norfolk Islands, should begin by acknowledging some of what it took to produce it—or at least some enumeration of the manifold generosities, obligations, tolerated impositions, kindnesses, and bonds of kinship that underwrote it.

There is, of course, the money. This book was built from work in archives around the world, and that in turn was underwritten by grants from numerous bodies. The Australian National University Center for European Studies awarded me a visiting fellowship, and the National Maritime Museum in Greenwich awarded me a Caird Short-Term Research Fellowship. The Princeton History Department funded, at considerable expense, travel to Pitcairn and Norfolk Islands. I received needed time and support from a fellowship at the Berlin Center for the History of Knowledge. There, Anke te Heesen, with incredible generosity, arranged support for me to travel to archives in London. At Denison, I received the Suzanne B. and Theodore A. Bosler '56 Endowed Faculty Fellowship, which helped support further research trips to archives in Australia and the United Kingdom.

There is, too, the necessity of acknowledging the academic venues in which parts of this work, or earlier iterations of it, have appeared. I have given talks on some aspect of this project at the University of Wuppertal Interdisciplinary Center for Science and Technology, the Australian National University Centre for European Studies, a conference titled Knowing through Collecting at the HU-Berlin Center for Anthropological Research on Museums and Heritage, the North American Conference of British Studies, and a superbly organized conference on Science and Islands in the Indo-Pacific World held at Cambridge. Some parts of chapters 2 and 3 were presented at the conference Discovering, Surveying, Ordering: Expeditions in the Long 19th Century at Helmut-Schmidt-Universität. Chapter 2 also benefited from presentations at the Midwest Conference on British Studies, the British Association for Victorian Studies, and the North American Victorian Studies Association. Chapter 3 was much improved by two conference panels: first as part of a panel I organized for the History of Science Society Annual Meeting, and then at the Pacific History Association. Versions of chapter 5 received valuable feedback from presentations at the HU-Berlin Colloquium in the history of science, the History of Science Society, and the European Society for Science and Technology Studies. Together, the comments, questions, comments disguised as questions, and questions disguised as comments offered by co-panelists, chairs, commentators, and audiences made this work better.

Some earlier versions of this work have also appeared in print. Elements of chapter 3 were published in 2022 as an article in *History and Anthropology* titled "Putting Anthropology in Its (Hospitable) Place: Harry Shapiro's Fieldwork on Pitcairn Island, 1934–1935." Additionally, some of the thinking on marginality and derangement in this book was presaged in a book chapter coauthored with Alexander Mawyer, Tillman Nechtman, and Joshua Nash, titled "Oltremari' dell'Europa d'Oltremare: Pitcairn, Mangareva e le Persistenti Interazioni Tra Isole Dimenticate [The overseas of overseas Europe: Pitcairn Island, Mangareva and the persistence of interaction between forgotten islands]," in Adiano Favole's edited volume *L'Europa d'Oltremare* [Overseas Europe], published by Raffaello Cortina Editore in 2020. An expanded English-language version, coauthored by the above as well as with Maria Amoamo, Martin Gibbs, and Pauline Reynolds, appeared as "When Margins Are Centres: De-Ranging Pitcairn Island's Place in Pacific Scholarship" in the *Journal of the Polynesian Society* (now *Waka Kuaka*) in 2021. I have tremendous gratitude to my coauthors and anonymous reviewers for their help in sharpening my thinking.

Like so much scholarly work, this book is the product of several institutions—and the people who made intellectual life come alive in those places. At Princeton, it was shaped by almost too many people to enumerate. Among my fellow graduate students, thanks go to Anthony Acciavati, Elaine Ayers, Michael Barany, Edna Bonhomme, Alex Chase-Levenson, Henry Cowles, Will Deringer, Margarita Fajardo, Joshua Garrett-Davis, Ben Gross, Heidi Hausse, Even Hepler-Smith, Cynthia Houng,

Rebecca Johnson, Emily Kern, Paul Kreitman, Meg Leja, Iwa Nawrocki, Ingrid Ockert, Andrea Onate-Madrazo, Flori Pierri, Ray Thornton, Iain Watts, and Nurfadzilah Yahaya. I also thank my teachers and mentors, especially David Cannadine, Linda Colley, Angela Creager, Michael Gordin, and Erika Milam. Above all, my thanks go to Graham Burnett, for introducing me not only to the scholarship with which this book engages but the humanistic modes that inform all of my work. I also want to thank my Denison colleagues, among them Lauren Araiza, Jessica Burch, Can Dalyan, Adam Davis, Leslie Hempson, Rebecca Kennedy, Tess Lanzarotta, Lisa Morrison, Julie Mujic, Trey Proctor, Mitchell Snay, Miranda Sachs, Mariana Saavedra-Espinosa, Hosna Sheikholeslami, John Soderberg, Karen Spierling, Joanna T. Tague, Megan Thelkeld, Shao-yun Yang, and Hoda Yousef. This book benefited from their collective wisdom. It also benefited from the very helpful work of two student research assistants, who helped me with processing archival material: Emma Borgia and Teddy Stahl.

I have accrued many debts in the wider world of academe, too. I had instructive conversations with Warwick Anderson, Chris Ballard, Bronwyn Douglas, Margaret Jolly, Peter Mühlhäusler, and so many other scholars in Australia, all of whom left their marks on this book. Rosana Dent, Adam Fulton Johnson, Joanna Radin, and Isaiah Wilner helped me to think about the histories of fieldwork, indigeneity, and hospitality. Meg Davidson generously shared archival sources with me, and Herb Ford very hospitably welcomed me at the Pitcairn Islands Study Center. Michael Eshleman kindly discussed sources in the legal history of Pitcairn Island. Maria Amoamo, Martin Gibbs, Alexander Mawyer, Joshua Nash, Tillman Nechtman, and Pauline Reynolds have been generous coauthors who have helped me think through so much regarding Pitcairn and Norfolk Islands. I would not be a historian of science or of colonialism were it not for Jim Bartholomew and Alice Conklin. Riki Kuklick welcomed me into the world of the history of anthropology as a young graduate student, and in that way shaped this book, too. Finally, I especially thank Emma Ching at the University of Hawai'i Press and the two anonymous reviewers of this book—their generous critique has made it considerably better.

I have other debts closer to home. My thanks go to my family for dealing with me at my most busy and stressed: Kevin and Mary Alice Snyder, Jon and Debbie Rausch, AJ and Domy Mata, David Snyder. I thank Mike Young for a lifetime of support and encouragement. I thank Angelika Young for the same, and for bringing decades of copyediting expertise to many of my drafts. Mary Snyder I thank for everything.

However, my ultimate acknowledgments go to Pitcairn and Norfolk Islands and their diasporas. Thanks to all those who hosted me and spoke with me on or about both islands, including but by no means limited to Anne Harrison, Lisa Richards, Albert Buffet, Alice Buffett, Colleen Crane, Tom Lloyd, Trish Magri, Jodie Williams, Jay Warren, Carol Warren, Kevin Young, Melva Evans, Meralda Warren, Vaine Warren-Peu, Charlene Warren, Heather Young, and Kerry Young. Thanks, too, to those with whom I spoke but promised not to quote. Thanks, above all, for hosting me, another in a long line of outside researchers visiting to produce yet another in a long line of books about your histories, your communities, and your homes. I suspect it will not reveal much that you do not already know, and I suspect it will not be the last. I hope that I have treated your islands and your histories with the respect they deserve. And I hope, above all, that it changes the ways some of the researchers who follow me portray your islands and other communities like them.

Introduction

Making the Natural Laboratory

> On Pitcairn, we see an experiment in culture contact—the impact of European or English, civilization and Polynesian. It occurs, to be sure, on a small scale, but it is a reduction in degree only. Here we have a number of advantages for the study of phenomenon. The situation has been simplified to its very elements. Most of the imponderables which affect other foci of culture contact and the complex reactions that obscure the design are absent on Pitcairn.[1]

So wrote American anthropologist Harry L. Shapiro in 1936. The people he described, Pitcairn Islanders, were already famous among scientists like him. Descended from eighteenth-century British sailors and Pacific Islanders from Tahiti, Tubuai, Huahine, and Raiatea, they seemed to constitute a uniquely "hybrid" society that, by virtue of their supposed isolation, had developed into a useful microcosm of the wider world. On Pitcairn, Shapiro suggested, an accident of history had stripped away complicating factors, and even so inscrutable and complex a phenomenon as culture contact could be "simplified to its very elements." The island had become a place where truth about human nature was laid bare. It was, in other words, a natural experiment.

The notion that islands, and especially Pacific Islands, can serve as ready-made experiments or human laboratories has persisted at least since the early nineteenth century. Their smaller size and separation from the mainland have made them seem bounded and scrutable, sites where extraneous factors have been stripped away to make fundamental truths more visible.[2] Few insular communities have been described as more laboratory-like than Pitcairn Island, a one-by-two-mile volcanic outcrop 1,200 miles west of Rapa Nui (Easter Island) and 330 miles east of the Gambier Islands, inhabited at its height by fewer than 240 people. Over the last two hundred years, Pitcairn Island and then Norfolk Island, another insular space to which much of the population of Pitcairn was relocated by the British Empire in 1856, have played host to hundreds of writers and researchers.

Many of them construed both islands as ready-made experiments, model systems, or social worlds in miniature, places where knowledge about human beings was easier to discern.

And yet, those who conducted research on Pitcairn and Norfolk Islands often found very little in the way of laboratory-like control and scrutability. Rather, visitors often grew anxious that long histories of colonial meddling and constant investigation had instead produced only a hospitable facade, one either consciously or unconsciously constructed to hide real truths from the prying scrutiny of outsiders. Surveying a half century of government reports and investigations in 1910, for instance, the British colonial governor Lord Chelmsford declared that "there is no place where it is more difficult to find out the real truth than Norfolk Island."[3] The same was often said of Pitcairn; Albert Wadkins Moverley, their onetime schoolteacher and the first linguist ever to study their language systematically, wrote in 1953 that "the hallmark of the successful Pitcairner is the ability to 'hypocrite the stranger,'" an act so common that the community had supposedly evolved its own word to describe it.[4]

This book is a history of these two contradictory and yet deeply interwoven myths: on the one hand, that Pitcairn and Norfolk Islands could serve as natural laboratories making visible fundamental truths; and on the other, that both islands are places of heightened mystery or even ostensible duplicity, shielding their inscrutable truths from view. To trace the emergence and trajectories of these intertwined notions, I follow two centuries of interest in both islands, investigating the work of their investigators. Pitcairn and Norfolk's scrutinizers came from knowledge-making communities of many kinds, including sailors, colonial bureaucrats, anthropologists, linguists, and historians. Together, these writers and researchers produced a sizable body of knowledge. Surveying the vast bibliography on Pitcairn alone before justifying his own manuscript about the island in 1964, New Zealand photographer and adventurer Hardwicke Knight estimated that "some 2,500 historical, scientific, and romantic books and articles have been published on various aspects of the subject."[5] His was likely an overestimation, but the number is large and has only grown since. Indeed, on a per capita basis, the Pitcairn Islanders certainly rank among the most written-about people ever to have lived.

That persistent interest has come with costs. Though alluring and rhetorically useful for those who deployed it, the metaphor of the "natural laboratory" often failed in practice. Pitcairn and Norfolk were never, their visitors found, as isolated or as microcosmic as they had imagined—and as one visitor followed another, they often grew concerned that long histories of scrutiny and meddling by observers like themselves had only further eroded the islands' laboratory-like distinctiveness. Far more crucially, the strange endurance of the natural laboratory myth has resulted in an invasive and recurring scrutiny for Pitcairn and Norfolk Islanders themselves. For two centuries, outsiders have again and again examined every detail of their lives, deploying and redeploying descriptions of them in print to bolster their own theories and careers. Pitcairn and Norfolk Islanders appeared frequently in racialized and eugenic visions of both Britain's Pacific empire and the human sciences, subject to exoticizing and sexualizing depictions that centered their "moral" and reproductive lives. Those visions not only were the result of colonial structures, but also contributed to further colonial interventions, among them coerced resettlement projects meant to maintain the Pitcairn Islanders as racialized experimental subjects. Throughout that two-century history, Pitcairn and Norfolk Islanders worked to contest those depictions and interventions, and to pursue their own autonomous visions—and too often these were ignored or marginalized by those who would study or govern them. Accordingly, this book is above all a historical critique of the myth of the island natural laboratory, an idea that emerged with colonialism, proliferated in and beyond the Pacific, and persisted with particular acuteness in studies of Pitcairn and Norfolk Islanders.

THE MUTINY ON THE *BOUNTY*

That anyone should take an interest in Pitcairn Islanders at all depended on a maritime legend well known to the Anglophone readers and filmgoers of the last two centuries: the mutiny on the *Bounty*. As a phrase, "the mutiny on the *Bounty*" was once enough, by itself, to conjure thoughts of imperial ambition, romantic rebellion, and exoticized Pacific life. Though its status has slowly eroded these last decades, the story was once canonical in English-speaking culture.[6] When Britain's National Maritime Museum held a bicentennial exhibit on the *Bounty* in 1989, its catalog professed that "nearly everyone has heard of Bligh, his mutiny, and his epic voyage."[7] Accounts began with attestations of the mutiny's universal recognition well before then, too. In 1922, the author of a short story about the South Pacific wrote, in what reads as both declaration and exhortation, that "everybody must know the story of the mutiny of the HMS *Bounty*."[8] A half century before that, an 1871 article in a Boston literary magazine asserted that "the main facts of the *Bounty* mutiny are no doubt familiar to most of our readers."[9]

The readers of this book might know the story, too, from a variety of sources that retain some prominence. The most famous telling in print was a trilogy of novels published between 1932 and 1934 by Charles Nordhoff and James Norman Hall, the last of which centers on Pitcairn itself.[10] Their books eventually sold some twenty-five million copies and found readers across the globe.[11] One visitor to Pitcairn recalled obsessing over the novels as a small boy in 1930s South Africa, writing that "like millions of other boys of my age in the English-speaking world I was entranced by this, the greatest sea story of all time."[12] Directors and movie studios retold the story, too, in major motion pictures starring Errol Flynn, Clark Gable, Marlon Brando, and Mel Gibson.[13] And, of course, there were history books, including mainstays such as Greg Dening's and Caroline Alexander's.[14] In many of these, Pitcairn Island is a last chapter or an epilogue, a space of denouements where all becomes clear and plots find their resolutions. I am not a historian of the *Bounty*, and this book is not a history of the mutiny. Nevertheless, a knowledge of what transpired during and after that event, as well as some understanding of its erstwhile cultural prominence are necessary to make sense of the investigations into Pitcairn and Norfolk Islanders' lives that followed.

While historians disagree considerably in their interpretations of the mutiny's causes and consequences, most agree on a general narrative. In October of 1787, the Admiralty dispatched Her Majesty's Armed Vessel *Bounty* to the Pacific under the command of Lieutenant William Bligh. His mission was to procure breadfruit saplings, which the naturalist Joseph Banks hoped would serve as a cheap foodstuff for the enslaved people of Britain's Caribbean colonies. The *Bounty* arrived at Tahiti after a year's journey in October of 1788. Its crew spent five months anchored at Matavai Bay, gathering trees and entangling their lives with those of Tahitians, a complex cross-cultural encounter that most histories recount simply as one of hedonistic sexual hospitality. On April 5, 1789, the *Bounty* weighed anchor and began the long voyage home. A few weeks later, on the morning of April 28, 1789, master's mate Fletcher Christian led a mutiny, taking over the *Bounty* and setting Bligh and some of his loyal crewmembers adrift in the ship's launch; the deposed captain would lead them on an epic journey to the Dutch colony of Timor. The mutineers sailed back to Tahiti, where some chose to remain. Most, however, feared the prospect of reprisal if found, and sought a more permanent home where they could live out their lives in Pacific obscurity.[15] They attempted to settle on the island of Tubuai, but gave up after their violent colonization attempt resulted in the murder of dozens of its people. They tried again. After abducting twelve women and six men, most of them Tahitians but some from Tubuai and Raiatea, Christian and eight other mutineers sailed away from Tahiti for the last time.[16] They tacked eastward, toward Pitcairn's Island, recently discovered for Europeans by Philip Carteret in 1767 and described in his account as uninhabited and ringed by cliffs.[17] They arrived on January 23, 1790. After running their ship aground, they salvaged what they could before burning it down to the waterline. Far above the cliffs, they built a small settlement, which would later be christened Adamstown.

Most histories describe the island's early years as a period of racial and sexual conflict.[18] The English mutineers parceled out land among themselves, granting none to anyone else. The mutineers each took a partner, leaving a remainder of three women for the six Tahitian men. Most authors agree that after the partner of one of the mutineers died, things broke down. The Tahitian men killed five Englishmen, including Christian, in a surprise attack. The surviving mutineers fled into the bush and, with the help of some of their partners, proceeded to kill the Tahitian men. Soon, the remaining mutineers turned on each other; according to most accounts, one was murdered with an axe to the head, and another committed suicide by tying a stone around his neck and leaping from a cliff. When the killings were finally done, two Britons survived, along with most of the women and their children. One was Edward "Ned" Young, who died of "asthma" in 1800. That left Alexander Smith, also known as John Adams, the last mutineer on Pitcairn. He lived until 1829.[19] Most historians rely on his accounts when narrating the island's early years, though sometimes these are buttressed with accounts by the island's founding women, whose voices, agency, and historical authority, as I show in a later chapter, were too often neglected in *Bounty* and Pitcairn histories.[20] Indeed, most historians retell these early years as a minor episode of idealized colonialism in which British men founded and shaped a new settlement in *terra nova*. However, if we look beyond this framing, the founding of Adamstown in 1790 is perhaps better understood as a second arrival of Pacific Islanders in a place already colonized at least once before, part of a long history of voyaging in a Sea of Islands.[21] Indeed, that historians have for so long sought Britishness and neglected Pacific lives and perspectives in their accounts is a consequence of the discourses and obsessions I examine in the chapters to come.

SCIENCE, COLONIALISM, AND THEIR DERANGEMENTS

This book, however, is not a history of the *Bounty*, or of Pitcairn, or even of Pitcairn Islanders—though they are an active and crucial part of the story it tells. Rather, what I offer here is a history of the ways outsiders and strangers (including myself) have studied and written about Pitcairn

Islanders for their own ends, often with considerable disregard for the concerns or voices of those about whom they wrote.[22] There have certainly been many strangers. Outsiders encountered the Pitcairners for the first time in 1808, inaugurating a series of visits and producing two centuries of discourse, much of it centered on the belief that the Pitcairn Islanders' communities could serve as useful microcosms for the wider world.

Strangers' interest had many consequences, one of which was sporadic but powerful colonial intervention. Britain's formal claim over Pitcairn Island is usually said to extend at least back to 1838.[23] Even now, the UK continues to govern the island as an external territory; it remains, in a literal sense, the reason the sun does not set on the British Empire. Imperial rule was haphazard, however. With rare exceptions, no British administrator lived on the island, and it was only occasionally visited by passing warships. And yet British colonial rule also could be frightfully powerful. On two occasions, the imperial state initiated coerced migrations: first an abortive attempt to transplant the community to Tahiti in 1831, and then a more permanent move to Norfolk Island in 1856. Beginning in 1859, a few families back-migrated to Pitcairn, and there have been permanent Pitcairn-descended communities on both Pitcairn and Norfolk Islands ever since, as well as a diaspora in New Zealand, Australia, and elsewhere. Over the decades, colonial rule has remained a force on Norfolk Island; the Australian government, which came to administer it as an external territory, revoked the island's capacities for self-rule in 2015.[24] Colonial governance has also returned sporadically on Pitcairn, including a police investigation and a series of trials for child sexual abuse at the beginning of the millennium, which in turn led to a reimposition of British authority.[25]

Pitcairn and Norfolk Islands' history as subjects of colonial scrutiny is not unrelated to their history as objects of study, and especially to their use as "models," "natural experiments," and "natural laboratories."[26] Historians and literary scholars have already shown that these tropes emerged within imperial contexts, especially in the case of islands.[27] That was the case for Pitcairn and Norfolk, too, where cross-traffic between science and colonial administration was considerable. However, beyond revealing obvious linkages between agents of knowledge and agents of power at any given moment, the examples of Pitcairn and Norfolk show us that the "natural laboratory" concept emerged slowly and historically, dependent on a series of iterative interactions between investigators and investigated over generations. Historians of science have not attended very much to iteration in field sites like these, and our histories are often not deep enough. Books and articles centered on specific figures or expeditions by necessity are often unable to examine rigorously the ways field sites are embedded in longer histories of repeated encounter. We also fail, sometimes, to recognize scientific visits as only a subset of other arrivals and departures. In the case of Pitcairn, it was sailors, romantics, and evangelical writers who first imaged the community as exemplary or even sacred, a stage sequestered from the wider world that could offer truths about morality and human nature. Imperial administrators, seeking to protect that image of a Pacific idyll, deliberately scientized the islanders through a forced colonial relocation it regarded as a racialized "experiment" in colonial settlement, deploying the language of science not as metaphor but as description of colonial practice. Only later did scientists and historians take up Pitcairn and Norfolk as ready-made cases. An attention to the long, slow emergence and uses of Pitcairn and Norfolk Islands as places scrutinized by nonscientists thus shows us more than the context in which science was done—it shows us the colonial origin of notions like the island as human laboratory.

At the same time, *longue durée* attention to sites like Pitcairn and Norfolk also shows us persisting anxieties about the boundaries of knowledge and power.[28] Remarkably, even though these were places celebrated for reducing the problems of the world to simplicity and clarity, researchers repeatedly asserted that truths there were actually more difficult to ascertain, or even impossible to determine. Indeed, investigators often experienced both communities as sites of epistemic derangement, places where they found their attempts to produce knowledge halted, redirected, or reconfigured. I use the term "derangement" very intentionally here, building on Ann Stoler's observation that in colonial archives, "epistemic uncertainties repeatedly unsettled the imperial conceit that all was in order."[29] I do not at all mean to suggest that the islands or their communities were themselves deranged or that they mutinously refused the gaze of their studiers, though that was an accusation often leveled at them. Rather, I use the term to describe the ways that the projects of knowledge makers maddeningly broke down, often resulting in substantial reconstructions of core categories or assumptions, as disorderings gave rise to reorderings.[30]

In many ways, these epistemic derangements resulted from limitations and paradoxes implicit in the notion of insular metonymy or the natural laboratory itself. For one thing, they were hardly "natural" or "accidental" laboratories; both Pitcairn and Norfolk were deliberately experimentalized not just by the colonial gaze but by deployments of colonial power. For another, the language of the "laboratory" or "experiment" was often deployed less as a description of practice and more as a rhetorical mechanism for legitimating or justifying knowledge claims.

Writers often used such disparate terms as "laboratory," "experiment," "model," "microcosm," or "exemplary space" as near synonyms, and rarely rigorously defined them. These images appeared frequently in published books and articles, but almost never in private field notes; on the islands themselves, investigators learned quickly that Pitcairn and Norfolk offered nothing like the control they imagined was afforded to scientists performing experiments in literal laboratories.[31]

Finally, and perhaps most crucially, the idea of the island as natural laboratory was built on conceptions of spatial or even metaphysical separation from the world, which of course proved impossible to maintain in practice. Many writers described both Pitcairn and Norfolk Islands as utopias or, in a counter-discourse, as secret dystopias.[32] In Thomas More's original formulation, however, "Utopia" was a play on words, meaning not only "good place" but also "no place." Utopia was impossible, a fantasy conjured in order to critique the worldly here and now. Pitcairn and Norfolk were depicted as similarly fantastical and separate, as it was their separateness that made them useful as metrics or mirrors of the real. Foucault called such sites *heterotopias:* "other places" simultaneously disconnected from and contiguous with the larger world, places "neither here nor there" in which society re-creates or rediscovers itself.[33] Pitcairn and Norfolk Islands were broadly understood to be *both* here and there, part of and apart from the rest of the world all at once. Epistemic derangements resulted when researchers attempted to close that impossible distance.

To "derange," definitionally and etymologically, means to move from order, to displace, to disarray.[34] In imaginings of Pitcairn or Norfolk Islands as separated from the world and as real spaces one could visit, in conceptions of them as profoundly different and didactically similar to the mainland, the islands were de-ranged or re-ranged, dis-placed and re-placed across space and the imagination, dragged closer to or further from the metropolitan center to suit an author's needs. It is worth remembering that, though Pitcairn Island was often described as one of the world's most distant places, marginality is not a fact of geography, but is always a product of perspective and imagination.[35] A place can be marginal only relative to another place. Historically, Europe imagined the Pacific as an extreme colonial periphery, while for much of the Pacific world, islands were "navels" from which the rest of the globe appeared as varyingly distant margins in their own right.[36] The notion of Pitcairn or Norfolk as distant elsewheres was always a deliberate construction. Indeed, even insularity itself was a manufactured notion in some respects; as Sujit Sivasundaram has shown, islands could be understood as connected to mainlands until the British Empire "islanded" them.[37] I suggest that visitors to Pitcairn or Norfolk often failed to find the natural laboratories they sought because the metaphor itself inhabited the impossible boundary between the real and the imagined, and investigations collapsed it. On closer inspection, researchers often recognized that the islands they imagined as isolated experiments were in fact not isolated at all. Both were always connected to other islands, sometimes receiving hundreds of visits a year.[38]

The recurring result of these disjunctures was epistemic derangement, the throwing into disarray of the categories under study. Much fieldwork has been described as frustrating or disillusioning, to be sure.[39] But because Pitcairn and Norfolk were understood as places where cross-cultural, cross-racial, or cross-linguistic contact played out on a microcosmic stage, knowledge makers used them to elucidate questions of mechanism, hoping the products and processes of contact would be more discernible than usual. In the 1920s and 1930s, anthropologist Harry Shapiro visited both islands to understand how racial traits were inherited across generations. In the 1950s and 1960s, Pacific historian Henry Evans Maude studied Pitcairn in order to produce an exportable model of colonial resettlement. And from the 1950s through the end of the century, an array of linguists attempted to study the ways that languages merged in order to form new, hybrid forms. And yet, all came away from their encounters with the islands questioning the categories or assumptions that drew them to those sites in the first place. What was race, and did it even exist as a biological reality? What was historical authority, and could it legitimize postcolonial governance? What were the boundaries or definitions of a language in the first place? All that seemed epistemologically solid melted into air.

ISLANDER AGENCY AND ITERATIVE INVESTIGATION

Because Pitcairn and Norfolk Islands, like so many Pacific Islands, were repeatedly subject to investigation, researchers encountered the lingering detritus and accumulated debris of those who preceded them, even as they added more layers and entanglements themselves. It is a vision of the field site that runs counter to Western assumptions about the practice of science; even an earlier generation of critical science studies scholars imagined knowledge as a product extracted from peripheries and collected in what Bruno Latour called "centers of accumulation."[40] However, as much-studied places, sites like Pitcairn and Norfolk Islands became centers of extraordinary accumulation in their own right, places where visits, observers, and observations piled up in tremendous abundance. No one accumulated more knowledge about or experience with investigations on Pitcairn and Norfolk Islands than Pitcairn and Norfolk Islanders themselves. They were and remain subject to sometimes hundreds of investigations across their lifetimes (including, it must be said, my own)—an example of science's "overstudied others."[41]

Though this is largely a book about colonial officials and scientists from Europe or settler societies who sought to study and use Pitcairn Islanders for their own purposes, the people they investigated were not the passive objects of study their investigators so often desired. Though cast as scientific objects, they emerge again and again in the archive as speaking subjects. Pitcairn and Norfolk Islanders are part of the fabric of the story that unfolds in the chapters to come, shaping the knowledge made about their communities. They hospitably hosted their guests, guiding them through the pounding waves that ring Pitcairn and lodging them in their homes—modes of reception that would shape the patterns of scientific work. They offered knowledge of their pasts and presents, not just as "informants" but as active entrants in published discourse.[42] Too often histories of science and empire deal in binaries between studiers and studied, but these broke down from the beginning, as Pitcairn and Norfolk Islanders authored accounts in print, and some earned PhDs and became researchers themselves. Perhaps most crucially of all, Pitcairn and Norfolk Islanders fought for both political and epistemic autonomy. Though often mediated, the voluminous archives of investigation on both islands are replete with moments in which the purported subjects of scrutiny spoke back on and about the colonial record, telling and shaping their own stories. Outsiders sought to stage the islands as compliant utopias and laboratories, but they often refused to be staged. Every chapter of this book will thus tell not only histories of outside investigation, but also of autochthonous contributions and contestations.

However, the story that unfolds across this book is also one that requires a nuanced and sensitive portrayal of Pitcairn and Norfolk Islanders' agentive roles. The archive of colonial scrutiny and scientific investigation is, after all, overflowing with concerns about "islander agency." Writers and visitors, hoping to find compliant experimental subjects, often worried that their interlocutors were *too* practiced as hosts, and had become too adept at performing in accordance with expectations. At the extremes, this emerged as a running paranoia that Pitcairn and Norfolk were actually places of artifice and duplicity, mutinously hiding dystopian truths beneath an Edenic veneer. And yet, this discourse should not suggest that such a facade in fact existed, even as a hidden transcript of resistance or agnotological strategy.[43] Rather, I suggest that it was largely a product of the concerns and obsessions of investigators themselves. Visitors sought an untouched Eden, but were also aware of long histories of previous research and colonial entanglement. Moreover, Indigenous secrecy has long been an obsession of anthropologists and other investigators like them; the ability to access hidden or secret truths made their own investigations and knowledges appear more valuable, by virtue of their having supposedly broken through and accessed something rarer and more real than their predecessors and colleagues.[44]

But perhaps the most remarkable feature of these iterative investigations and colonial anxieties has been their uncanny perdurance. Again and again, visitors found Pitcairn and Norfolk Islands very un-laboratory-like. Again and again, Pitcairn and Norfolk Islanders were not silent objects of discourse but active entrants in it, working to counter suggestions that their communities were utopic or dystopic microcosms. Even so, the myth of the natural laboratory would not die, and remains ubiquitous in Pacific scholarship even in this century; a 2013 edited volume surveying the Pacific justified study of the region by asserting that "the diversity, isolation, and small-scale dimensions of islands presents laboratory-like opportunities for research on general questions about nature and society."[45] Neither has the idea died on Pitcairn and Norfolk, where research relying on notions of isolation and laboratory-like scrutability remains ongoing.[46] Settler academia continues to forget the very unnatural histories of these natural experiments.

The rest of this book is an attempt to remember. Each of the next five chapters follows a set of knowledge makers, the knowledge they produced, and the derangements that arose in the course of their work. Chapter 1, on sailors, shows how a haven for mutineers was rehabilitated in the Anglophone imagination as a stage setting for morality plays and Edenic fantasies of Pacific settlement. That utopianism resulted in considerable meddling by outsiders, including interventions by the colonial state and bids for paternalist dictatorship by newly arrived strangers. Chapter 2, on colonial administrators, argues that efforts to transplant a Pacific Eden through coerced resettlement resulted in the deliberate experimentalization of Pitcairn Islanders as racialized subjects. Having established how these natural laboratories were made by colonial forces, the second half of the book follows the ways researchers took up Pitcairn and Norfolk as exemplary and experimentalized spaces. Chapter 3, on anthropologists, follows the work of Harry L. Shapiro on Norfolk and Pitcairn Islands. Expecting laboratory-like spaces in which the mechanics of racial inheritance would be as clear as in geneticists' jars of fruit flies, he encountered a community shaped by long histories of previous encounters and grew anxious that his informants could not be trusted. Chapter 4, on historians, shows the ways that Pitcairn Islander Rosalind Young and Pacific historian Henry Evans Maude used the islands' pasts in order to support visions of both historical and governmental authority in their respective presents, writing with or against the notion of the islands as colonial models. Chapter 5, on linguists, historicizes the ways researchers deployed the islands as "pristine" laboratories for working out the dynamics of languages born from cultural contact. Their work ultimately turned toward more fundamental questions, as they debated the definitions and boundaries of what constitutes a "language" in the first place.

Finally, a short conclusion does its eponymous work, summing things up and drawing out meanings. However, I am skeptical that this is a story that has in any way concluded, despite my ambition to help bring it to a close. I suspect I will not be the last writer to use Pitcairn and Norfolk Islanders' lives to tell stories for their own purposes. I know that the long-running interest in Pitcairn and Norfolk Islanders by outsiders like me is built on the deranged interest of those who preceded us, even as we sustain it further. Hardwicke Knight, the New Zealand adventurer who visited in 1964, was obsessed with Pitcairn well before he arrived; he had already read all that he could about it in some of the 2,500 sources he identified and, once on the island, spent much of his time attempting to retrace the steps of the explorers who came there before him.[47] He remained obsessed well after he left; he studied photographs of rock carvings on the island's cliffs again and again, retracing their lines on fresh sheets of paper. Most archaeologists agreed that these petroglyphs were left by Pacific people some eight centuries ago. Knight, however, saw in some of their geometric forms the shape of a European sailing ship, or perhaps even something stranger; his notes contain images comparing one outline to a menorah.[48] He showed University of Otago archaeologist Helen Leach a stone tool he thought had been deliberately flaked to form a map of the island.[49] Knight saw Pitcairn everywhere, and the world in Pitcairn. Friends worried that his obsessive interest was turning toward the eccentric, perhaps even a kind of madness.[50]

I see too much of myself in Knight. I have also read Pitcairn and Norfolk accounts by the hundreds, ever since I first encountered them as an undergraduate student in a biological anthropology course. I, too, have retraced the steps of those who preceded me, and I, too, have found that history of investigation deranging. As a doctoral student in the history of science, I visited Norfolk Island in 2014. Leafing carefully through the journals and correspondence of past researchers in the archives of the island museum, I was acutely aware of a small shelf of previous studies, written by my predecessors, lurking above me. In the British National

Figure 1. British artist Rhiannon Adam photographing tourists photographing Pitcairn Islanders (not pictured). Photograph courtesy of the author.

Archives, I found yellow request slips by authors whose names I recognized falling from the folios. And on Pitcairn, I retraced Knight's retracings, following the steps of past visitors on its dusty clay paths. That is always the case for historians, or any authors entering into a scholarly discourse; we all have our predecessors and we join conversations that precede us. But for those of us who study the relationships between knowledge and power in colonized spaces, the recognition that we, too, are caught up in past entanglements can be vexing.

I sometimes protest that I do not study the studied, but instead the studiers. And yet, I know that I am clearly among them. On Pitcairn, I remember meeting photographer Rhiannon Adam, there to produce her own account of the community; the island is so studied that sometimes visits by researchers overlap. I photographed her while she photographed German tourists who were themselves photographing Pitcairn Islanders (see figure 1).[51] The layers of observation and meta-observation piled up, but I am not sure they amount to any appreciable distance. I remain aware that my own ambitions are not altogether different from other knowledge makers who seek to produce their truths about the wider world using the Pitcairn and Norfolk communities. I surely have my own fantasies, anxieties, and preoccupations about the histories of these two insular spaces, and my vision is shaped by an awareness of the scholarly and colonial entanglements that preceded me. If anything, a desire to create distance between my own project and those of the historical subjects I scrutinize in the coming chapters might be among the most obvious of de-ranging ambitions, as I feverishly imagine a gulf where none exists. At best, it is my hope that a historical awareness of their and my own derangements can help Euro-American, continental, and settler academics and writers like myself to understand how the notion of the island society as human laboratory emerged, to think critically about why it has persisted so long, and to begin the work of dismantling it.

CHAPTER ONE

Eden Refounded

Sailors and the Construction of an Exemplary Space

The curtain drew open. Revealed before the audience was "a picturesque glen," flanked on each side by mountains. The waves of the Pacific lapped in the distance. A palm tree grew center stage. Amid the scenery was the better part of a theater company, dressed as Pacific "natives" and English sailors, presenting spectacular "Indian Sports" and a "Grand Ballet." This was the 1816 premiere of a new operetta, *Pitcairn's Island,* at the three-thousand-seat Theatre Royal on London's Drury Lane. Its titular setting was an island only recently "discovered" half a world away, and a long subtitle ensured that audiences recognized the performance's ripped-from-the-headlines appeal: *A New, Romantick, Operatick Ballet Spectacle Founded on the Recent Discovery of a Numerous Colony Formed by, and Descended from the Mutineers of the Bounty Frigate.*[1]

Though the operetta's subject and setting were borrowed from contemporary reports, *Pitcairn's Island*'s playwrights proved unconstrained by newspapers' descriptions—its plot was straightforward romance and adventure. Fletcher Christian's fictive daughters, Otaheina and Oberea, witness the arrival of a British contingent on the shores of their isolated paradise. It is a dramatic and perilous moment; Christian and his fellow mutineers are perhaps the most notorious criminals alive, men so desperately sought that the Admiralty dispatched the *Pandora* to the far end of the world to find them. Otaheina asks her sister: "How could you go / to lead me so / among these handsome strangers? I cannot stay; / unless we run / most surely are we all undone!"[2] But the sisters need not have worried. The British sailors prove as good-willed as they are handsome, singing "Britannia's legions only seek / against the strong to guard the weak."[3] True to their word, the sailors and the descendants of mutineers soon band together to defeat a party of hostile "natives." The island pacified, Oberea and Otaheina soon fall in love with two of the Britons, and Pitcairn, once the hidden den of the empire's most infamous scofflaws, is rehabilitated. Its British-Tahitian community has earned its place in the empire by fending off the "savage threat," and the last scene displays the "grateful attachment of the Islanders" as the navy departs.[4]

Pitcairn's Island debuted twenty-seven years after the *Bounty* mutiny, but only a few years after the island itself was "rediscovered" by the American whalers of the *Topaz* in 1808 and again by the Royal Navy vessels *Briton* and *Tagus* in 1814. Alongside accounts in newspapers and periodicals, the melodrama was one of the first images of the island and its inhabitants any Britons encountered. However, the questions it raised would be imposed on Pitcairn by outside visitors and writers for some time. Were the islanders, the descendants of traitorous mutineers, loyal and trustworthy? What was their identity, in any case: Were they "British," "Tahitian," or something new—something hybrid, perhaps even dangerous? Could attachments to Britain and its culture convert them into loyal subjects? In posing and offering answers to these queries, *Pitcairn's Island* helped to instantiate a discursive treatment of the Pacific island as a kind of stage, a literal one in this instance, on which to work out the nature of British identity and the future of Britain's Pacific empire. In so doing, the melodrama became an early entry in an ever-expanding library of texts that used the island as a microcosmic space for working out macrocosmic problems, especially those of empire and identity. The resulting idea, that Pitcairn could be exemplary, particularly in didactic or moral terms, would persist for centuries.

This chapter, which surveys writing about Pitcairn Island by outsiders in the first half century after its rediscovery, will unpack the early history of its staging as an exemplary space—as well as contestations to that staging on the part of Pitcairn Islanders themselves. As we will see, Pitcairn emerged in nineteenth-century British discourse as a site where the work of defining self and other was especially acute. The island played host to hundreds of visiting sailors, many of whom generated written descriptions dedicated to analysis of the islanders' identity and the meanings their island afforded. These texts emphasized the islanders' "Britishness" and celebrated Pitcairn as a bastion of surprising familiarity in an otherwise alien and distant "fifth part of the world."[5] However, though writers consistently treated the island as a model, the lessons they drew from it shifted. Early accounts offered romantic descriptions of Englishness transplanted to and successfully hybridized by the Pacific. By mid-century, however, outsiders increasingly promulgated an image of the island as a moral and religious Eden, deploying the Pitcairners as living examples of the prayerful life to which their readers should aspire. A wide Victorian readership came to know Pitcairn as a moral, Anglican utopia—a sacred paradise set apart from the profane world.

Many histories of Pacific islands have already treated their depiction by travelers as places freighted with cultural and epistemic weight.[6] Indeed, many historians and literary scholars located in texts about the Pacific not only the formation of the colonial subject, but the making of the European self.[7] As a space celebrated for its hybridity, however, Pitcairn offers a particularly distinctive vantage point from which to examine how writers and sailors used the Pacific to define themselves and their Britishness.[8] In much of the writing on identity and colonialism's culture, encounters with non-Western alterity came to define the Western self through acts of negation.[9] Pitcairn Islanders were sometimes described in the same mode—they were *more* literate, *more* handsome, and *more* Christian than the "Polynesian" inhabitants of the islands to their east, who in turn were accorded a higher ranking on racial hierarchies than other Pacific peoples. However, more often than not, descriptions of the Pitcairners emphasized their uncanny familiarity, as though they were English people transplanted and modified strangely, but not negatively, by the infusion of Pacific qualities—from the Pitcairners' uncanny "hybridity," writers and sailors produced a garden of fertile possibility. In their bodies and in their community, self met other and produced peaceful, loyal, and devout British subjects. In that way, writers and sailors construed them as symbols of the success of the missionary and settler projects, object lessons in imperial aspiration.

And yet, we can also locate the origins of long-lasting epistemic derangements in the accounts of sailors and those writers who drew from them. Etymologically and definitionally, to "derange" is not just to enter a state of madness, but to move from an orderly pattern or place; in positioning Pitcairn Island as simultaneously utopian and real, both profane and sacred, early nineteenth-century writers closed and opened impossible distances. By conjoining the island to and separating it from the world it was meant to exemplify, they de-ranged and re-ranged it, bringing it closer to or further from the metropole again and again in order to make it familiar or distant, real or fictive.[10] Sailors and the romantic authors they inspired, like many in the first decades of the nineteenth century, freely mixed romantic hyperbole and realist verisimilitude. Those texts produced a widely known legendarium that in turn shaped visits to come, as sailors increasingly expected to see Edenic text enacted in the flesh. At the same time, as those accounts accumulated both on and around Pitcairn, paranoid anxieties began to emerge that the island was not what it seemed, and that any real "truths" about its inhabitants lurked behind a facade constructed to conform to utopian expectations. The result of both discourses would be violent colonial intervention. Strangers sought not only to project but to protect their utopian fantasies, at times even forcibly staging them through impositions of autocratic control or coerced relocation. At the same time, as this discourse and its resulting colonial impositions emerged, Pitcairn Islanders themselves fought for self-representation, and for the capacity to enact their own autonomous futures.

ROMANCING ENGLISHNESS IN THE SOUTH SEAS

The island's first ever encounter with strangers has been told and retold hundreds of times, an example of the ways meetings between Pitcairn Islanders and sailors generated not just tellings but retellings in text. The island's own record, a one-volume register of important events, is perhaps the shortest recounting of the occasion: "1808. Arrived Ship *Topaz* of Boston, Folger Master."[11] Another, more widely published version is Captain Folger's recollection as retold by New England sailor Amasa Delano. His rendition foreshadowed encounters yet to come, as even in their very first meeting with outsiders, the Pitcairn Islanders were forced to submit to scrutiny over their identity.[12]

In Delano's account, the American whaling ship *Topaz* halted off the coast of a mis-charted island, and, as they prepared a landing party, a canoe approached. To the Americans' astonishment, its rowers hailed them in English. The captain answered them, introducing himself as Mayhew Folger, of Boston, in America. They replied: "You are an American; you come from America; where is America? Is it in Ireland?" Taken aback, Folger tried again to discern the islanders' identity:

FOLGER: Who are you?

PITCAIRNERS: We are English.

FOLGER: Where were you born?

PITCAIRNERS: On that Island which you see.

FOLGER: How then are you Englishmen, if you were born on that island, which the English do not own, and have never possessed?

PITCAIRNERS: We are Englishmen because our father was an Englishman.

FOLGER: Who is your father?

PITCAIRNERS: Aleck.

FOLGER: Who is Aleck?
PITCAIRNERS: Don't you know Aleck?
FOLGER: How should I know Aleck?
PITCAIRNERS: Well then, do you know Captain Bligh of the Bounty?[13]

That moment was revelatory. The story of the mutiny on the *Bounty* was already famous, and Folger knew it well; even at its very first encounter with strangers, Pitcairn Island was understood by outsiders as part of a larger maritime legend. Delano tells us that Folger felt a profound pleasure at the realization that he was witness to the living vestiges of a famous tale: "The whole story immediately burst upon his mind, and produced a shock of mingled feelings, surprise, wonder, and pleasure not to be described."[14] "Aleck" was "Alexander Smith," the last surviving mutineer of the *Bounty* on Pitcairn. The textual had been made suddenly, almost impossibly, real.

Folger wrote the Admiralty in 1809 to inform them of his discovery.[15] The British government never responded. Perhaps preoccupied with the prosecution of wars against France and then the United States, it regarded a lone missive by a Yankee whaler as only so much distracting rumor. Folger's correspondence was eventually published in the *Quarterly Review*, though the editor of that London publication noted that "if this interesting relation rested solely on the faith that is due to Americans, with whom, we say it with regret, truth is not always considered as a moral obligation, we should hesitate in giving it this publicity."[16] The tale was corroborated, however, in 1814, when Royal Navy vessels *Briton* and *Tagus* visited the island. Their descriptions of Pitcairn, along with those of Folger's, were published and republished in books and journals.[17] Soon, the story was everywhere, appearing in periodicals, poems, and theatrical spectacles.

One of the first to take up sailors' accounts and recast them for her own purposes was English poet Mary Russell Mitford.[18] In 1811, at the age of twenty-four, she published a long-form work, *Christina, the Maid of the South Seas*.[19] A narrative poem in four cantos, it imagined a romance between an English sailor and Fletcher Christian's daughter. It was not a classic by any means, and if it is remembered at all, it is only because Mitford eventually developed substantial acclaim as the author of *Our Village*, and because her poem prefigured Byron's own romantic telling of the *Bounty* mutiny and its aftermath.[20] Almost none of the voluminous literature around Pitcairn Island so much as acknowledges the existence of Mitford's poem. And yet, so far as I can determine, it was the first long-form written work ever published about the island. It is a remarkable initial entry in that corpus, one that was centered on definitions and critiques of Englishness, and that found generative possibilities in Pitcairn Island as a space overgrown with vibrant, fecund hybridity.

Christina, the Maid of the South Seas might strike us as mere romantic adventure. Its fictive plot certainly tends toward the formulaic, and it is populated almost entirely by invented characters. "Henry" is an English voyager crossing the Pacific on an American ship captained by his friend "Seymour." Their vessel is damaged in a gale, and its crew are forced to land on a remote island. There they find the lost mutineers, including handsome young "Hubert" and his betrothed "Christina," Fletcher Christian's daughter. The community is led by patriarchal "Fitzallan"—Fletcher had long ago leaped from a cliff in a fit of romantic inner torment.[21] Henry, wandering the island's woods one moonlit night, finds Christina mournfully attending her mother's funerary urn in a hidden glen. He watches her until she departs, and then picks up the tear-covered flowers she has dropped, falling, of course, hopelessly in love. Fitzallan, meanwhile, narrates the island's past for the visiting American sailors inside a majestic cavern, explaining that years ago the Tahitian men of the island rose up, killing all but him. Fletcher Christian's widow, in turn, poisoned and stabbed the murderous men. Now, conflict between Englishness and Tahitianness threatens to reassert itself as Hubert, playing romantic airs on his flute, enchants Christina and wins her heart. She is torn between the incestuous, brotherly affection of her affianced Hubert and the adventurous heart of the English sailor. Hubert and Henry nearly come to blows as the Americans make ready to leave the island. Christina goes sullenly to her wedding with Hubert, only to discover that her fiancé has reconciled with Henry and has given the couple his blessing. Love conquers all, racial tension is tamed, and Henry and Christina live happily ever after. The mutineers' haven becomes a loyal colony.

As in the *Operatick Spectacle* performed on Drury Lane, Mitford's Pitcairn became a stage for romantic passions—passions that were resolved when the island's heritage of mutiny was mollified by the next generation's romantic reattachment to Britain. Its romantic tropes were obvious: enchanted glens, hidden caves, turbulent suicides, star-crossed lovers, and so on. But Mitford's poem was not only an exercise in imagination. Rather, hers was the first in a long line of texts about the island to fuse romantic speculation with empirical description—an attempt to de-range, or close the impossible distance between reality and utopia. In her preface, Mitford claimed that "Fitzallan's Narrative, romantic and improbable as it appears, is entirely founded on facts."[22] She drew the premise itself from Folger's actual encounter with the islanders in 1808, which she claimed to have learned of through an anonymous "gentleman, who heard from several officers of the *Topaz* an account of the manners, the virtues, and the happiness, which she has attempted to portray."[23] Indeed, the entire text was meticulously researched. Endnotes totaling 140 pages annotated her work, substantiating detail after poetic detail with careful citations from nearly every prominent account of the Pacific then available.[24] She also reproduced the only published source about Pitcairn then in existence, Mayhew Folger's first letter to the Admiralty published in the *Quarterly Review*.[25]

Many of Mitford's citations served to substantiate small details about the islanders' way of life. She used a library of expeditionary accounts to shore up even two-line couplets; the note to her lines "The link'd ivahahs, side by side, / Short poles at once, unite, divide" included a lengthy excerpt from James Cook's first voyage from which the reader learned not only that the "*ivahah*" is a kind of canoe, but also its other names, classes, sizes, constructions, and uses in warfare.[26] The phrase "Oh generous people!" led to a lengthy disquisition, again from Cook, on gift giving.[27] Other footnotes offered descriptions of dress from Wallis, the exchange of names from Alvera Mendana de Neyra, and so on.[28] If the lines "from the mountain height / peers the wild goat in rude afright" were included in her poem, it was because Mitford read in Bligh's account that "among the articles which they brought off to the [*Bounty*] and offered for sale, were capsicoms [*sic*], pumpkins, and two young goats."[29] The lines "rose the gay fig" demanded an excerpt from the naturalists Joseph Banks and Daniel Solander.[30] Neither animal nor vegetable was left to the imagination.

Yet Mitford's poetic vision elsewhere was unconstrained. She admitted as much, changing the names of her actors for mostly aesthetic reasons. "Alexander Smith" was in her estimation "surely the most unpoetical appellation by which [a] hero ever was distinguished," so she rechristened him "Fitzallan."[31] Mayhew Folger became "Seymour." Other actors she invented entirely; Fletcher Christian never had a daughter Christina, neither was there a Hubert.[32] The island's houses became "extremely picturesque and beautiful" English cottages "resembling the better sort of those usually seen in Devonshire."[33] No English adventurer on the American *Topaz* ever romanced the island's most beautiful maiden, or at least no such romance is remembered elsewhere. Rather, Mitford dismissed her work as a flight of fancy in her closing lines: "Oh! It is sweet, in this disjointed age / To 'scape awhile life's sad realities." Her Pitcairn was, she suggested, nothing more than a romantic escape from life in Regency Britain.

But if it was only a flight of fancy, why did Mitford so meticulously research and annotate her text? Footnotes and endnotes were hardly unknown in Romantic poetry. William Wordsworth interspersed his poems with footnotes referencing geographies and histories; Samuel Taylor

Coleridge, a mentor of Mitford's, hemmed in the 1817 edition of "The Rime of the Ancient Mariner" with a running prose description; and Reginald Heber, whom Mitford herself cited and much admired, corroborated flowery descriptions of Levantine geography with citations from travel accounts in *Palestine* (1803). But Mitford's notes were even more exacting and more prolific. Footnotes can serve as defensive structures, meant to ward off potential attacks and shore up points of weakness; perhaps they served to protect a young woman entering a literary field in which her contemporaries were not, by most measures of social capital, her peers. Just as crucially, Mitford was well aware that both Bligh and the families of the better-connected mutineers, especially the Christians and the Heywoods, carefully guarded their reputations. Mitford noted in her preface that "it was difficult so to write, as to avoid on the one hand the charge of palliating a most fatal conspiracy and, on the other, an imputation far more dreaded by the Author!—of irritating the feelings of a highly respectable family, and tearing open the scarcely healed wounds of kindred affection."[34]

However, more was at work. In methodically fusing her poetry to a corpus of travel accounts, Mitford imbued her romantic flight of fancy with the weight and heft of the "real," attempting to close the impossible distance between Pitcairn as imagined utopia and Pitcairn as actual place. Thus began the de-ranging work of transforming Pitcairn into a heterotopic space, both connected to and disconnected from contemporary society—but metonymic and reflective of it at the same time. Travel narratives like Cook's were widely read as authoritative depictions of otherwise-inaccessible peoples and places, as were fantastical or imagined versions of them, such as Diderot's *Supplément au voyage de Bougainville.*[35] *Christina* was both. Small details about Pacific customs from Cook's or Philip Carteret's texts gave the appearance of substance to the otherwise insubstantial—and in consequence made Pitcairn into a living utopia, real but distant, useful for the work of assessing Mitford's own society.

Mitford dwelled considerably on the problems and possibilities of Pitcairn's hybridity. The landscape of Mitford's Pitcairn Island thrived on images of Englishness transplanted—and she mostly wrote of "Englishness" even if she perhaps imagined a wider Britishness. Pitcairn became a crossbred garden of Pacific and European cultivars, where "fruits of every clime unite / As if some fay, from Europe's shores; / Had borne them to that balmy air, / To bloom in fresher beauty there."[36] It was a beneficent mixture that was mirrored by its inhabitants. The island's fusion of Polynesian and English strains showed the most aesthetically pleasing virtues of each, though Britishness emerged as the dominant—if strengthened—form. Describing the second generation, she wrote: "The towering youth, the graceful maid, / Were both in Indian garb array'd; / but not a trace of Indian feature / Appear'd in either glorious creature."[37] Rather, they were representations of Englishness draped in the trappings of Tahitian culture, preserving the purity of the "noble savage" while maintaining the enlightened air of civilization: "Of polished life they own the sense / Savage alone in innocence!" It was, in fact, the combination of paradisiac purity and English industry that rendered Pitcairn such a unique and beautiful place in her imagination. As Fitzallan recounted: "A lovely desart [*sic*] we had found, / If desart 'twere, where all around / Liv'd plant, and flower, and flowering tree, / A silent world of faëry! / Soon felt the vale of British spade; / Soon rose the cottage in the shade."[38] The metaphor of transplantation was an apt one, relying as it did on the *Bounty*'s original mission to collect breadfruit saplings. However, it also proposed that Englishness could survive and even thrive in new and hybrid forms through its encounter with the Pacific world.[39] Her poem, like the Drury Lane operetta, suggested Pitcairn had transcended racial tension to become a successful colony, and that Britishness was resilient enough to do so in other spaces, too. From its first appearances in British texts, Pitcairn was construed as a model, separate from but representative of the wider world, useful for projecting visions of race, identity, and empire.

NAVAL VISITS AND INTERROGATIONS OVER IDENTITY

As ships began to visit the island with more regularity, more descriptions of it saw print. These narratives not only celebrated the community's unique origins but also maintained an interest in the islanders as transplanted and hybridized Britons, often informed by a racialized gaze that sought out and diminished any Pacific influences on the community. That was certainly the case for the second and third ships to visit Pitcairn, HMS *Briton* and HMS *Tagus*, which arrived in 1814 and resulted in several published accounts.[40] Captain Pipon of the *Tagus*, for instance, devoted much of his text to descriptions of the islanders themselves. The island's young men he deemed "finely formed, athletic and handsome." Its women were beautiful and surprisingly modest. They "have invariably a piece of Linnen reaching from the waist to the knees, & generally a Mantle or something of that nature thrown loosely over the shoulders & hanging as low as the ankles; this however is frequently thrown aside & often entirely off. . . . It is not possible to behold finer forms."[41] Nonetheless, as much as he indulged in the colonialist exoticism and eroticism that usually accompanied descriptions of Pacific Islanders, it was the presence of Englishness that made the islanders especially noteworthy. Pipon depicted Thursday October Christian, Fletcher Christian's son, as dressed in only a loincloth, a poncho around his shoulders, and a hat bedecked with feathers. But his manner was described as friendly, his English was reported as grammatical, and the officers were "glad to trace in his benevolent countenance all the features of an honest English face."[42] For Pipon, morally and racially inflected signs of Englishness were crucial above all.

Lieutenant H. B. Willis, during the same 1814 encounter, also underscored Pitcairn's Englishness as morally and racially uncanny. In a journal entry, he recorded utter astonishment at seeing the small community of English speakers, who possessed "the most athletic and finest forms, exhibiting the most beautiful faces and the most admirable symmetry and shape." They formed a "society of human beings very unlike their wretched progenitors. Blessed with innocence, harmony, and peace, though scantly provided, [they] exemplify how few things are really essential, in any situation, to make a community, when pure, comfortable, happy, and independent."[43] Already, writers were drawing moral lessons, expressed in the hierarchical language of race and civilization. Willis illustrated his journal entry with a full-color rendering of Pitcairn's green peaks and rugged cliffs, flanked on both sides by human figures. These blended ethnographic with classical style; he cast their young healthful bodies with European features and draped their bodies with light, flowing cloths. In keeping with other European accounts of the Pacific during this period, "Polynesian" peoples were regarded as more akin to Europeans on imagined racial hierarchies—but Willis saw an even closer kinship.

Among the most romantically tinged early descriptions of Pitcairn Island is that of John Lawrie, a sailor who visited the island on the *Hercules* in 1822. British colonists in Calcutta, having read about the island's "rediscovery," purchased a shipment of gifts and supplies for the community, which the *Hercules* delivered. Though replete with ethnographic descriptions of Pacific peoples, Lawrie's journal is mostly devoted to discovery of a more introspective variety. He had joined a ship bound for the South Seas after a love affair turned sour; "deprived of every hope, and sickened with every pleasure, I am forced to fly into a foreign land, there to seek shelter from all agony, and await the lenient hand of time to sooth my woe."[44] Claiming inspiration from Robert Burns, he recounted his experience on the island in a poetic register. "Mysterious Isle!—so long the seat of fear / When England searched for 'Christian' and his peers—/ How strange to find so sweet an Eden here!" he wrote after seeing Pitcairn for the first time.[45] Lawrie, too, agreed that the islanders were a simple but beautiful people, living peaceful lives beyond the problems of the wider world. Theirs was a community that, during his brief visit, he found deeply affecting. On leaving Pitcairn, he wrote: "Farewell, sweet ocean speck, a long farewell! / Yet, wakeful as the waves that round thee swell / My soul shall watch thee, as a thing Devine / *apart from the world,* from vice and crime."[46] Notably, it is the uncanniness of Pitcairn's existence as a utopian space, in but separate from the real world, that sustained his interest.

Captain Frederick Beechey's account of his visit with HMS *Blossom* in 1825 was among the most read and retold, though his stay produced texts by some of his officers, too.[47] All of them evince a sense that the island was already widely celebrated. The narratives of previous naval visitors to Pitcairn had circulated through the Anglophone world, and were read by other sailors. George Peard, a lieutenant on the *Blossom,* recorded in his journal that "every one's curiosity was raised to the highest pitch to behold a spot which had afforded shelter to part of the mutinous crew of the Bounty, and been colonized by them." He wrote, "Some were even look[ing] forward to find him, Fletcher Christian."[48] This remained a commonplace fantasy, though of course most accounts already reported the leader of the mutiny dead.[49] Beechey recorded the crew's anticipation, too: "The interest which was excited by the announcement of Pitcairn Island from the mast-head brought every person upon deck," he recalled, "and produced a train of reflections that momentarily increased our anxiety to communicate with its inhabitants; to see and partake of the pleasures of their little domestic circle; and to learn from them the particulars of every transaction connected with the fate of the Bounty."[50]

As the *Blossom* closed in on the island, a canoe ventured out to greet it. This time, the last surviving mutineer was on board. His name, he admitted, was not Alexander Smith at all, but John Adams. It was a suitable surname for the patriarch of a new Eden. He came dressed in the remnants of his old uniform and "retained his sailor's gait, doffing his hat and smoothing down his bald forehead whenever he was addressed by the officers."[51] Mirroring the reattachment to empire that pervaded the last decade's poems and operettas, Adams and the younger Pitcairn men eagerly shook the hands of every officer with "undisguised feelings of gratification."[52] They invited the ship's crew ashore, and soon a party of officers and men were rowing through the dangerous surf at the landing place, which Beechey christened "Bounty Bay" (see figure 2). The name, which indelibly inscribed the island's romanticized past on its landscape, has endured.

The *Blossom*'s crew assured Adams that he would not be arrested for mutiny. Amnesty came at a cost, however. Beechey and his officers interrogated him about his role in the insurrection, the fate of the mutineers,

Figure 2. Landing in Bounty Bay, c. 1830. Oil painting. Attributed to Richard Brydges Beechey, based on a drawing by Frederick William Beechey. Mitchell Library, State Library of New South Wales, ML 114.

and Pitcairn's past. They pressed the old mutineer for details, compiling them into a seven-page written testament to which Adams, barely literate, signed his name.[53] Adams's narratives were widely republished and helped establish an early history of the island as one of masculine competition: English men defeated Tahitian men, and the Tahitian "wives" were won over and converted, loyal and obedient. However, they also ironically made Adams famous as a paragon of piety and loyalty. Meditating, like Mitford before him, on the island's paradisiac power to convert its inhabitants from evil to good, the captain wrote, "It would have been wonderful, after the many dreadful scenes at which they had assisted, if the solitude and tranquility that ensued had not disposed them to repentance."[54]

In the rehabilitation of the last mutineer, displays of Anglican devotion were especially effective, and Adams was not subtle in demonstrating Pitcairn's religiosity. He orthodoxly adhered to the practice of saying grace before every meal. When Adams accidently took a bite of bread mid-conversation, he suddenly "recollected himself, and feeling as if he had committed a crime, immediately put away what he had in his mouth, and commenced his prayer."[55] The *Blossom*'s officers visited the island on a Sunday, during which he led a church service. Adams was sure to include prayers for the royal family, offered "with much apparent loyalty and sincerity." Other prayers "were added to the usual service; and Adams, fearful of leaving out any essential part, read in addition all those prayers which are intended only as substitutes for others." Another islander then delivered a sermon, and, "lest any part of it should be forgotten or escape attention, it was read three times." Hymns were sung, first by the adults and then again by the children. The entire service was repeated at sunset. With the addition of separate morning, midday, and evening prayers, Beechey tallied that "they may be said to have church five times."[56]

Performances of religion were a means by which Pitcairn's residents were thought to display their Englishness and their patriarch's redemption. Beechey and his crew, however, also looked for signs of otherness lurking beneath demonstrations of Englishness. His principal descriptive register was the romantic; an amateur artist from a family of painters, he produced dramatic sketches of the island with exaggerated precipices and canoes charging through towering breakers—staging a geography of cliff-bound separation from the wider world. He romanticized the island's women, too, taking note of Pitcairn's "pretty black-eyed damsels" and their innocent glances in the standard, colonialist mode of South Pacific eroticism.[57] His officers searched out signifiers of "Tahitianness" in their way of life; both Lieutenant Peard and Captain Beechey recorded that the islanders baked their food by burying it next to heated stones, and noted proscriptions against both genders eating meals together.[58] They also sought literal performance of Tahitian heritage. One night, after dinner, one of them asked for a dance. The Pitcairners were reluctant, telling the sailors that by custom they refrained from dancing.[59] The officers insisted, and three women relented, agreeing to perform.[60] Beechey described "a reluctance which showed it was done only to oblige us," and described the dance as "little more than a shuffling of the feet, sliding past each other, and snapping their fingers."[61] Moments like this abound in accounts like Beechey's. "Englishness" and "Tahitianness" appear and are read as performances, and often only as metrics for assessing status on imagined racial hierarchies. Lost in the colonialist description was the perspective of the dancers themselves, and what these cultural forms meant to them. The result here, and across accounts of Pitcairn Islanders, was instead a masculine-centered gaze that perennially read signs of Tahitianness as feminized and obscure, and that compelled the performances it wished to see.

Emblematic of that gaze was scrutiny over race, and especially the scientization of Pitcairn Islanders' bodies. Beechey's officers were among the first to search out signs of alterity anatomically, investigations of racial status that prefigured the island's later history of anthropometric investigation. Alexander Colie, the ship's surgeon, measured the circumference of their skulls and recorded their skin and hair color.[62] Already, other markers of identity were imposed on Pitcairn. However, in defiance of the racialist taxonomizing such as the kind imposed by Colie's amateur phrenology, the islanders contested any categorization of themselves as "black." Their narratives of the island's early history, at least those they told their British guests, were founded on English victory over Tahitian influence, a division they maintained during their conversations with outsiders, and which elided the history of their mothers. John Shillibeer, during the visit of the *Briton* and *Tagus,* recorded that the islanders said Christian was "shot by a black fellow," and wrote that "the hatred of these people to the blacks is strongly rooted, and doubtless owes its origin to the early quarrels which Christian and his followers had with the Otaheiteans after their arrival at Pitcairn's." His account included an anecdote, often cited in later literature, in which Thursday October Christian encountered a West Indian sailor. "Christian looked at him sternly, rose, asked for his hat, and said, 'I don't like that black fellow, I must go,' and it required some little persuasion before he would again resume his seat."[63]

Racial descriptions appear throughout Beechey's account, but he assigned the islanders to no discrete category. They were "South Seas" in some respects, "English" in others—though he did compare them to other Pacific Islanders more often than his contemporaries. Above all, however, Beechey unvaryingly portrayed them as endearing. As the landing party made preparations to return to the *Blossom,* the community gathered to offer its farewell and departing gifts. It was a gray, foggy day, and the ship sailed slowly away from Pitcairn as the islanders' distant cheers grew inaudible. Beechey wrote that he was "unconscious until the moment of separation of the warm interest their situation and good conduct had created in us."[64] Like other visitors, he inscribed his lingering affection for the Pitcairners in his travel narrative. Closing his account, he declared: "All which remains to be said of these excellent people is that they appear to live together in perfect harmony and contentment; to be virtuous, religious, cheerful, and hospitable, beyond the limits of prudence; to be patterns of conjugal and parental affection."[65] His narrative would prompt further treatments of Pitcairn Island in print, and would drive further investigations and interventions in the community's lives by the colonial state—many of which occupied uneasy boundaries between racialist othering and romantic affinity.

FASHIONING AN ANGLICAN PARADISE

However, it was neither the islanders' professions of Englishness nor their racial status that most captured the Victorian imagination as the decades wore on. The island's romantic and racial valences never disappeared; Pitcairn persisted in the Anglophone imaginary as an idealized stage setting for the display of mutinous passions or romantic spectacles, and readers the world over remained captivated by the legend of the mutiny and the myth of the lost paradise. But by mid-century, writers increasingly fused those visions with explicitly religious and moral tones. Pitcairn's mutinous past became a necessary antecedent for a utopian present—an act of original sin had, in their view, ironically cast Adams's descendants into Eden. Writers began to understand Pitcairn not only as a romantic stage in which to work out problems of racial or national identity, but as a didactic object meant to demonstrate lessons in virtue. By the middle of the century, its inhabitants' Englishness increasingly became defined by their displays of Anglican devotion, as religious writing did a great deal to fashion Pitcairn Island into an explicitly exemplary space meant to demonstrate moral truths.

One of the more powerful men to become interested in Pitcairn Island as morally exemplary was Sir John Barrow, second secretary to the Admiralty. An admirer of Joseph Banks and a strong advocate for exploration, Barrow ordered several Royal Navy ships to visit Pitcairn, particularly Beechey's *Blossom*.[66] He was the anonymous author of many early articles about the island in the *Quarterly Review*, and likely the figure who published Folger's letter announcing the island's rediscovery.[67] In 1831, he produced one of the first full-length histories of the mutiny and Pitcairn Island.[68] In it, Barrow collected nearly every account then available, along with testimony from the mutiny itself, between which his own prose served as a kind of moral scaffolding. Barrow declared that "many useful and salutary lessons of conduct may be drawn from this eventful history." The first was the stark example of the mutineers' fates: insurrectionists and murderers got their comeuppance. A second was borne from what he saw as Pitcairn's founding irony: that from a genesis in sin, the islanders had produced an earthly Eden. Repentance and redemption were always possible. Barrow thought it "impossible not to feel a deep interest in the welfare of this little society," and in the coming decades, Barrow's Royal Navy captains would bear out that interest as the island's protectors and stewards, and he would publish their reports.[69]

At the same time, depictions of Pitcairn Island began to appear in children's texts. In 1827, a primary school reader excerpted an account from

Pipon's narrative of the *Tagus* voyage, from which pupils learned "that for morals, politeness of behavior, and . . . a strict adherence to truth and the principles of religion," the islanders had no equals.[70] After 1831, many didactic texts borrowed directly from the sailors' accounts excerpted in Barrow's history. An 1838 *Lecture to Young Men on Chastity* used Pitcairn Islanders to illustrate "the effect of a simple and correct regimen on the physical and moral character of youth."[71] They maintained a simple diet, exercised often, and abstained from all temptation; the health of their bodies reflected the health of their souls. An entry in the 1849 *Stories for Summer Days and Winter Nights* included "The Ship and the Island," which retold the tale within the frame of a grandfather narrating a seafaring adventure story to his grandchildren.[72] It borrowed liberally from texts like Barrow's to portray the Pitcairners as a chaste people with excellent moral and hygienic habits.[73] Tracts staged Pitcairn as a kind of parable, one entry in a string of moral examples of right living.

In time, entire books devoted to Pitcairn's edifying example appeared. One of the first was N. W. Fiske's hagiographic *Aleck: The Last of the Mutineers, or the History of Pitcairn's Island* in 1845.[74] Fiske took pains to assure his readership that the story of Pitcairn Island was "actual history," unlike those "fabulous tales which . . . have an unhappy tendency, however excellent their morality may be, to create and foster that love for novel-reading which has proved the ruin of multitudes."[75] From him, young readers learned that the mutineers got their just deserts and that their faithful descendants were eventually rewarded by God. In a late chapter, Fiske offers a conversation between a mother and her children in which they reflect on the tale they have just heard.[76] William, ever the diligent pupil, takes note of the story's first and most obvious moral lesson:

> WILLIAM: How little the mutineers gained by their crime!
>
> CHARLES: But God brought some good out of it.
>
> MOTHER: What do you mean by that, Charles?
>
> CHARLES: Why, if the mutiny had not taken place, there would not have been that happy little colony on the island; and was it not God that caused it to turn out so?
>
> MOTHER: Yes. The wickedness of Christian and his associates was overruled by God, so that much good has resulted. A fertile island, capable of supporting many inhabitants, which had for ages been a desert, has received a thriving colony. Here an interesting community is established in which industry and good order prevail, and the influence of religion is universally felt. It is to be hoped that many generations will enjoy these blessings. It is also a remarkable fact, that several persons, before thoughtless, have been awakened, by what they saw and heard in visiting this

> island, to ask "what shall we do to be saved," and have become hopefully pious.
>
> JANE: And, mamma, I think the story also shows us what good it does to teach children the Bible.
>
> MOTHER: Most certainly. The village of Pitcairn shows to every one, how desirable and useful it is for families and neighborhoods, that the children should study and regard the Bible. To what is all the order, industry, cheerfulness, thrift and happiness of that village owing? Suppose that Aleck, when the widows and their orphan children were left to his care, had allowed them to live just as they might please, without making the Bible their guide and rule; what would have been the consequences? Do you think that they would have become the lovely and harmonious society which has been described?[77]

Pitcairn neatly became an exaggerated object lesson in protestant, Victorian values. "Aleck" (Adams), onetime mutineer, was a redeemed and even sanctified figure. But the expressions of redemption mattered, too—racial tension gave way not only to Anglican piousness, but also to colonial industriousness as a "desert" island became an industrious paradise.

One of the more popular authors to explore the island as a religious example was Anglican clergyman Thomas Boyles Murray, a member of the Society for Promoting Christian Knowledge. His *Pitcairn, the Island, the People, and the Pastor,* first published in 1854 but reprinted in multiple editions across the nineteenth century, became a much-cited text.[78] Like others before him, Murray felt "wonder and gratitude to contemplate so exemplary a race, sprung from so guilty a stock," and prayed for the island: "May you long continue a living model of all that is lovely, and of good report; and may nations not disdain to follow your example!"[79] Worrying, like Mitford, that readers would regard the Pitcairn story as "too delightful to be real," Murray closed the impossible distance between reality and utopia by buttressing his account of the island with excerpts from a half century of travel accounts, including those of Shillibeer, Beechey, and dozens who followed; Captain William Waldegrave's 1830 visit with HMS *Seringipatam* became a familiar source.[80]

Like Barrow and Fiske, Murray sanctified the figure of John Adams. Here, too, he was the last mutineer, redeemed for his crimes by converting women and raising a generation of pious children. But in Murray's account, divine intervention was more direct. Sometime after the last murders, Adams dreamed, echoing a famous passage in *Robinson Crusoe.* In one vision, a horrible being appeared and threatened to stab him with a dart. In another, he saw the future hellscape to which he, as a sinner, was surely doomed. Murray offered these two episodes as the work of the Holy

Spirit, "whose merciful design it was to give [Adams] a better knowledge of himself, and a sense of the justice and goodness of God, and to bring him, an humble suppliant, to the throne of grace."[81] His hagiographic rendering of Adams lifted Pitcairn into the highest moral register, and it maintained that elevated status in religious and missionary tracts for years to come.[82] Indeed, so sanctified was Adams that visitors to Pitcairn began to collect his relics. A Victorian visitor to the museum of the Royal United Services Institution in London, for instance, could see a lock of his hair in an ornate gold frame, a kind of protestant reliquary placed next to other relics of the nation, including Horatio Nelson's coat and the *Victory*'s flag.[83]

By the middle part of the century, dozens of ships might visit the island in a single year. Sailors steeped in the literature surrounding Pitcairn expected nothing less than a sentimental and moral paradise. Fortescue Moresby, the son of Admiral Sir Fairfax Moresby, who commanded the *Portland* during its 1852 visit, recorded his anticipation before landing in Bounty Bay: "Having read so much about the mutiny of the Bounty, and the subsequent romantic history of the mutineers, which has resulted in the formation of a colony celebrated for their virtue, and simplicity, and religion, I experienced a feeling of something (I know not what to call it) on approaching the island, that I have felt when visiting some spot held sacred either from history or from being the scene of some Biblical relation; it is a secret kind of satisfaction."[84] He had already gleaned from poets and historians one of Pitcairn's more important characteristics; it was bounded off and held apart from the rest of the world not only by its insularity but by its exemplarity, existing at a physical but also a spiritual remove from the wider world. Accordingly, visitors like him came to Pitcairn expecting something altogether more sacrosanct than the banal places from which they had come.[85]

DERANGING UTOPIA

Early nineteenth-century narratives generated a vision of Pitcairn as an Edenic space, separate from yet representative of the wider world. Subsequent visits were, in a sense, attempts to close that impossible gap, and to realize dreams of Utopia. However, some outsiders were not content merely to project these fantasies, and instead intervened, materially and even violently, in order to enact them. Some strangers became would-be rulers, while others—especially powerful British colonial administrators—promoted the coerced relocation of the entire community to other islands. These attempts were "deranged" in the sense that they de-ranged the island, literally moving the Pitcairn community across the Pacific, as well as figuratively from the utopian to the worldly. However, I call these attempts "derangements" for other reasons, too. For one thing, they often emerged from or generated profound epistemological or political frustrations when experience failed to live up to utopian expectation. For another, they constituted colonialist projections that generated maddening consequences for the Pitcairn Islanders themselves.

To begin with, let us examine one example of de-rangement in print. In 1851, a *Fraser's Magazine* article took a hypothetical Pitcairn Islander on a violent flight of fancy:

> If you could suddenly whisk up into the air one of the descendants of the mutineers in Pitcairn's Island and suspend him over the high road in front of the southern entrance to the Crystal Palace, so that he might command a bird's-eye view of the great thoroughfare both ways, and all over the Park, and down through the glass road into the interior of the exhibition, taking in a panorama dense with population, heaving with movement, the probability is, that the man would either rub his eyes and imagine he was dreaming, or shut them again, lest in the stunning effect upon his senses, his brain might get bewildered.[86]

The Crystal Palace was an enormous iron-and-glass structure built in London's Hyde Park as the central venue of the Great Exhibition, a massive display of industrial and artisanal products from all over the globe. In many respects, it was the first world's fair, meant to demonstrate the heights British civilization had attained.[87] To measure the advancement of history, however, required an independent yardstick, and this author used the Pitcairners, drawn from the same "civilization" but sequestered from it, to provide the necessary standard. "It is, therefore," continued the article, "that we should like to blindfold a Pitcairn native, and snatching him away from his potato patch in a balloon, suddenly take off his bandages at that point over the road where the whole panting mass might be revealed to him at a glance. The exhibition itself would yield nothing half so suggestive as his astonishment."[88] It was a deranged notion, built on an impossible, heterotopic paradox: the idea that a place could be at once separate from and a part of the world.

That paradox collapsed into banality or disappointment when that impossible distance was closed. In 1881, the fantasy of a Pitcairn Islander looking out at the bustle of London became embodied reality when Pitcairn Islander Russell McCoy visited. McCoy's time in the imperial metropolis excited considerable interest, due in no small part to the work of enterprising stage managers who contrived to put him on display at the Royal Aquarium in Westminster, where visitors were invited, for a small fee, to meet this "veritable Pitcairn Islander."[89] But McCoy was a difficult figure to exoticize, and newspapers seemed to recognize the ironic disconnect between romanticized image and embodied reality. "There will be nothing but friendly welcome for this English South Seas islander, who comes guarded by the not ineffective talisman of his wife's wedding ring," declared one.[90] McCoy spent an afternoon on display before a sympathetic Anglican clergyman encouraged him to leave the stage.[91] The living instantiation of the fantasy was, as drama, unsatisfying. Russell McCoy was happy to answer questions about his great-grandparents, though he was mostly concerned about his wedding ring, which was broken and needed repair.[92] How his exhibition rated as entertainment in comparison to other exoticizing performances of non-European others the aquarium hosted is not recorded in any discernible sources. Instead, British reportage about McCoy reveals above all a visitor whose own perspective and ambitions stood apart from those who sought to exploit his story, or even the clergymen who would seek to "protect" him.

These same impulses to relocate and stage would be enacted by the British colonial state. Beechey, after his visit in 1825, reported concerns on the part of John Adams (Alexander Smith) that the growing community might outstrip its resources and would need to move elsewhere. John Barrow, who had long taken an interest in the island's affairs and imagined it as a kind of favored, British Eden, promoted the idea. Above all, he turned the machinery of the empire toward "the preserving of this fine race of men" in order to "preserve the same moral and religious character which created the lively interest in the minds of our captains."[93] British discourse concerning the relocation of the Pitcairners often centered on preserving the island's singular "moral" character, and these concerns shaped the British management of the migration. While several potential sites were offered, eventually the Pitcairners secured permission from Pōmare IV to build a new settlement on Tahiti. Some of the community's founders were from the island, and all had relatives there. However, in the racist and colonialist visions of British observers, this was also cause for some concern. One missionary wrote of the Pitcairners' move that "they really are an interesting people, but I'm afraid their morals will soon be corrupted by the Otaheitians."[94] The reciprocal was also thought to be true, however, as figures such as Barrow imagined the islanders as useful missionaries, a Christianizing and Anglicizing influence. Indeed the move was the first of several imaginings on the part observers who hoped the islanders, having already recreated Eden on Pitcairn, might reenact that re-creation yet again.

In March of 1831, HMS *Comet* arrived at Pitcairn to relocate the entire population.[95] Its captain, A. A. Sandilands, reported that the community deliberated for three days before all agreed to the move—though, as historian H. E. Maude has suggested, the decision was made under considerable pressure.[96] They arrived at Tahiti three weeks later at a moment of political conflict, but Pōmare IV hosted the community and ultimately provided the promised land to build new homes and new lives. Old family relations were renewed; Sandilands reported that one new arrival even found a sister still alive, whom she had not seen in four decades.[97] However, arrival in Tahiti also brought exposure to new pathogens, and disease spread rapidly. Faced with astounding hardship from sickness, the community sought a return to Pitcairn, which they secured on the US whaler *Charles Daggett*. Some Pitcairners explained their rationale in a letter penned to the ship's owners: "Great mortality has prevailed among us since we resided at Tahiti. We have buried twelve and we fear that should we remain here our number will soon be very small."[98] Five more would die after their return to Pitcairn, meaning that seventeen people out of a population of eighty-seven ultimately perished to enact Britain's colonial experiment in relocation. As the next chapter makes clear, it would not be the last such venture in coerced resettlement.

Even so, the Pitcairn Islanders' missive survives as a remarkable example of their own voices during this episode. So much of their perspective disappeared in the early colonial record. Often, the only figures who populated outsiders' accounts of Pitcairn Islanders were British-born "patriarchs" like John Adams, leading to an erasure of Pacific perspectives, as the archive instead privileged a masculine and British image of the Pitcairn community.[99] An example is Sandilands's list of passengers aboard *Comet* during the migration to Tahiti, which was broken down in three groups: men, women, and "Otahetian women"; only the last group required, in his mind, a qualifying adjective.[100] When describing Pitcairn Islanders as a whole, writers built on a discourse that sought out and reified Englishness, defined them by their forefathers rather than foremothers—they were usually described as the "descendants of the Bounty's Crew," or the "descendants of the mutineers."[101] Accordingly, though we know that arrival on Tahiti marked a return to an island where bonds of kinship were extensive, this is only gestured at in the colonial record.

However, the decision to return to Pitcairn itself is an act that cuts across the archive as a kind of refusal to be staged, an agentive derangement of the colonialist and missionary project to reshape the Pacific according to British designs. In that way, it was one of the first instances in which outsiders' ambitions to stage the islanders for their own purposes would disintegrate into failure. It also resulted in early contributions to a problematic and recurring discourse about Pitcairn Islanders that described actual or suspected acts of agency using the language of disillusionment or duplicity. In 1833, Royal Navy captain Charles Fremantle visited the restored community on Pitcairn and reported that through their contact with other Pacific Islanders, they had "lost much of that simplicity of character which has been observed in them by former visitors," and he worried that "if much left to themselves, and visited by many ships . . . they will lose what simplicity they have left, and will partake of the character of their neighbors the Otaheitans." He also claimed that "it was a most difficult matter to obtain the truth on any point which told at all to their prejudice; and it was only by cross-questioning them that I could arrive at it."[102] Eight years later, the medical officer on the HMS *Curacoa* described the community as "anxious to conceal the facts" of their private disagreements, "believing that it was only the character of their being a virtuous and innocent family which made the English Government, as well as the English people, take such an interest in their welfare and happenings."[103] Worries that the island's celebrated piousness and exemplary morality were only facades erected to comport to expectations or obfuscate darker secrets would remain a paranoid feature of writing about Pitcairn Islanders ever since. However, I suggest that we should not read these texts as archives of artifice—to do so would be to replicate colonialist visions of the island as mutinous. Rather, they constitute inaugural entries in a racist, colonialist discourse of anxiety and frustration, one that emerged from the detritus of iterative encounter and the failures of projected utopian visions.

This same discourse of supposed vulnerability and disillusionment resulted not only in forced migrations but also in a steady stream of outsiders looking to join or lead Pitcairn's imagined utopia. Like the migration to Tahiti, these were often described as efforts to "preserve" the island as a model community apart from the world, even if, by definition, these constituted impingements from outside. This discourse arose almost from the moment of Pitcairn's rediscovery: an 1820 article published in a British sailors' magazine urged some sailor or missionary to join the community and take the aging John Adams's place as leader.[104] Such a figure landed with the *Cyrus* in 1823. John Adams reportedly asked if any educated crewmember would take his place as the island's schoolteacher and moral leader, and John Buffett, an English sailor, took up the offer. Or at least that is how Buffett himself told it in a short memoir published two decades later in a Hawaiian newspaper.[105] According to his own telling, Buffett had traveled the world as a sailor; he was shipwrecked off Quebec, survived a typhoon off Manila, and now sought a quieter and more meaningful life. Before his 1823 arrival on Pitcairn, he had already read several accounts of the island, including Delano's, and "thought it the most eligible place in the world, a place free from temptation, and with no hindrance to prevent a man becoming Christian." He would make the island a stage for his own redemption. Another sailor on the *Cyrus*, John Evans, hid himself on the island until the ship departed, and he too was allowed to remain. They became the first of many "outsiders" to live on Pitcairn.

Buffett's influence lasted until 1828, when another pair of "outsiders" arrived in a ragged, unnamed vessel. Its captain, Noah Bunker, was seriously ill and asked to remain on the island; Adams and the islanders reluctantly gave him permission. Soon afterward, Bunker flung himself from the cliffs that ringed Pitcairn. The islanders found him on the rocks below with several broken bones, somehow still alive. A visiting whaler gave him laudanum for his pain; when no one was watching, Bunker drank the entire bottle.[106] Bunker was survived by his traveling companion, George Hunn Nobbs. Like Buffett before him, Nobbs offered a romantic curriculum vitae, and claimed that he had deliberately sought out Pitcairn as an Edenic refuge from the rest of the world in which to find redemption. He introduced himself as the "unacknowledged son of Francis Rawdon, Marquis of Hastings, and Jemima French, daughter of an Irish baronet who fled after the Irish Revolution."[107] His dying mother purportedly urged him to find some quiet spot where the family's transgressions could be atoned for. Nobbs mentioned Pitcairn as one such refuge, explaining that "much of its history . . . had come to my knowledge." She seized on the idea; her purported last words were, "Go to Pitcairn Island, my son, dwell there, and may the blessing of God rest upon you!"[108] Nobbs founded a rival school to Buffett's and began reading his own sermons on Sundays. His arrival disturbed some of the islanders' distant observers, who worried that this outsider would disorder paradise. "The seeds of discord are already sown," wrote John Barrow, who decided "this Nobbs is probably one of those half-witted persons who fancy they have received a *call* to preach nonsense."[109]

Nevertheless, Nobbs's influence lasted through the relocation to Tahiti and until 1833. That year, another Englishman arrived. He introduced himself as Joshua Hill and claimed he had been sent by London to serve as the island's governor. Like his predecessors, Hill introduced himself as a romantic adventurer ready to reform, and be reformed by, the island. He listed a litany of supposed honors and achievements: he had dined with princes and Bonapartes, commanded Royal Navy vessels, joined the Royal Society, traveled to North America and India, and had once maintained the most fashionable carriage in Paris. And, like his predecessors, he worked to secure authority. He designated several of the island's men as his "councilors" and named three youths as "cadets."[110] He ordered that islanders cease all communication with Buffett and Nobbs, and he forbade locals from marrying outsiders. When Buffett continued opposing his rule, Hill decided to make an example out of him, sentencing his rival to be hung by his hands in the church and flogged.[111] Worried for their lives, the island's other Englishmen fled as exiles on the next ship. Hill's rule would last until 1838, when he finally fell out of favor with Barrow and the navy.[112] The HMS *Imogene* was sent to remove him from power. Buffet and Nobbs returned, and the latter resumed his position as patriarch.

Figures like Buffett, Nobbs, and Hill are hardly unknown in the history of the Pacific. Beachcombers appear often in early accounts of Pacific islands, living out liminal existences on the margins of empires.[113] Hill especially is usually described as a petty conman, even if a biography by Tillman Nechtman has suggested we should take his rule and ambitions much more seriously as visions of British colonialism in the nineteenth-century Pacific.[114] I do not disagree, though I do suggest that we should understand the arrivals and interventions of these figures as extreme instantiations of a deranging impulse usually limited to textual depictions. On stages, in poems, and across Sunday school lessons, writers drew from sailors' accounts to reimagine Pitcairn Island as a space in which they could define themselves and their Englishness, in which they could define the future of Britain's Pacific empire, and in which they could model stories of moral redemption. Buffett, Nobbs, and Hill, with their dramatic autobiographical pronouncements, did the same. They transformed the island into a living text, their own stage set—deranging it with considerable disregard for those whose lives they disrupted.

They would not be the last. The rest of this book will follow later figures who saw Pitcairn and then Norfolk Islands as exemplary places, microcosms in which to resolve or understand macrocosmic problems—and who would interfere with the community in order to preserve their favored fantasies. Their efforts relied on the image of Pitcairn Island and its inhabitants generated by sailors, their narratives, and their readers in the first half of the nineteenth century. That work had already wrought tremendous transformations. A community of mutineers and Pacific Islanders became a locus for understanding the boundaries of Britishness, where the mixture of "races" and cultures had produced a colony that portended optimistic possibilities for Britain's empire in the Pacific. Over time, sailors and missionaries also reimagined Pitcairn as a site of Christian redemption and moral righteousness. The result was a growing belief that Pitcairn was an island beyond profane history, occupying an almost sacral realm, able to throw into relief earthly problems. And yet, there always lurked an anxious undercurrent, a lingering suspicion that a re-created

Eden was fragile or impossible, and that only the right imperial meddling could preserve it. And, always, there remained occluded by the colonial archive the stories and perspectives of the Pitcairn Islanders themselves, whose lives were upended by attempts to enact or preserve that constructed discourse, but whose agentive decisions could thwart imperial directive. The next chapter will examine the further history of such colonial interventions in the second half of the century, their consequences, and Pitcairn Islanders' responses to them.

CHAPTER TWO

The "Experiment"

Colonial Administrators and Eugenic Anxiety

In 1852, the governor of Van Diemen's Land, Sir William Denison, received a dispatch from the secretary of state for the Colonies. It instructed him to advise on a bold resettlement scheme: Was it possible to move the whole population of Pitcairn Island across four thousand miles of Pacific Ocean to Norfolk Island? Pitcairn was famously small, but its population was growing, and as in 1831, observers worried that a crisis of sustainability loomed. Unlike 1831, when the community migrated briefly to Tahiti, the islanders' allies in British naval and missionary circles sought a new home that might mimic Pitcairn's famous isolation. Norfolk, similarly "remote" but considerably larger, had long hosted a penal colony notorious for its brutality. However, the Crown was to close the prison, and the colonial office wondered whether the soon-to-be-abandoned isle could prove a suitable venue for the Pitcairners. The stakes, as they conceived them, were high. "You must be too well acquainted with the history of these interesting people to render it necessary that I should enlarge the subject here," the dispatch read; "it is enough to say that they are a community who, in the small and remote island which they occupy, have preserved an innocence of life and a gentleness and benevolence of manners which have gained for them the esteem, the admiration, and the goodwill of all by whom they have been visited."[1]

For many British naval officers, it was a tantalizing idea. A ship's surgeon visiting the penal colony on Norfolk Island in 1852 had heard the "rumour that the Gov't intended removing the convicts entirely and supplying their place with the Pitcairn Islanders," noting that "this indeed would be a change for the better, and convert what now might be considered a den of crime into a perfect paradise of a place."[2] The islanders' admirers celebrated the migration as a modern-day exodus, a colonial resettlement they understood in expressly moral or even biblical terms. By the middle of the nineteenth century, writers and missionaries had already transformed Pitcairn Island into a sanctified space; readers imagined the island as a living utopia, home to mutineers who, through an original sin of their own, ironically found themselves cast *into* a redemptive paradise. The proposal to relocate them was a chance to refound Eden yet again. It was also, from the British perspective, a test of antipodean colonial settlement on a small scale: Would the Pitcairn Islanders replicate their exemplary history, and would a community founded by Pacific Islanders and British criminals find success once more?

In 1856, the colonial state made its vision reality, coercively relocating the entire population to Norfolk Island, though some families eventually back-migrated and refounded the settlement on Pitcairn. Across the second half of the nineteenth century, the supposed success or failure of their resettlement became a running question in the Victorian imagination, taken up by administrators and scientists alike. In government reports, the public press, and scientific discourse, writers came to refer to the move as "the experiment."

What follows is a history of that "experiment," as it was understood by its operators, from its instantiation in the 1850s to its assessment across the next six decades. This chapter will survey the intellectual context of its formulation, historicize the criteria by which its success was determined and, ultimately, recover something of the experience of those who lived under scrutiny within its bounds and who worked actively to contest the discourse imposed upon them. Just as chapter 1 did in the case of sailors, the present chapter will take up colonial administrators as a set of investigators concerned with the production, use, and dissemination of "knowledge" about Pitcairn Islanders. Like other groups of knowledge makers, colonial administrators were engaged in a project of elucidating "truths" about the distant island people for whom they increasingly made themselves responsible. Their surveillance prompted a shift in the Anglophone world's conception of Pitcairn and then Norfolk Islands, reconfiguring them from romanticized paradises into racialized and scientized spaces. Indeed, the notion that Pitcairn or Norfolk could serve as living experiments or natural laboratories, so crucial to the investigations of twentieth-century scientists, was borne in no small measure from the work and obsessions of nineteenth-century bureaucrats. The myth of the natural laboratory on Pitcairn and Norfolk Islands stemmed in large part from the practices and visions of colonial bureaucracy.

However, while the administrative reconfiguration of Pitcairn Islanders into subjects of "the experiment" was a vital precondition for their islands' later uses as supposed natural laboratories, colonial administrators investigated Pitcairn Islanders for their own reasons. They came to regard Norfolk Island as a Pacific crucible purpose-built to test whether new settlements founded, even in part, by Pacific Islanders could "thrive" and, concomitantly, to test the viability of a colonial project founded on what they understood as racial improvement. However, while the founders of the "experiment" initially described the Pitcairners in utopian and religious terms, across the coming decades their successors increasingly focused their vision on matters of sex, racial degeneration, and eugenics. Ultimately, they abandoned visions of "racial improvement" in favor of "racial hygiene," especially as they became obsessed with eugenic concerns.[3] Concomitantly, colonial officials increasingly became paranoid that dystopian secrets lurked almost imperceptibly beneath a utopian veneer designed to thwart outside inquiry. That epistemic anxiety, emerging awkwardly and paradoxically alongside notions that the island was an epistemically legible experimental space, would carry forward into similar uses and derangements on the part of other knowledge makers across the twentieth century.

Finally, and most crucially, in the history of this colonial "experiment," we can locate both the severe effects of governmental intervention in Pacific Islanders' lives and moments in which that intervention was contested. Outsiders' visions of and obsessions with Pitcairn Islanders not only resulted in a second coerced relocation of the population in the space of twenty-five years, but also prompted even closer scrutiny of their "moral" affairs. In the decades and centuries that followed "the experiment," the community would see invasive investigations into their lives, bodies, and reproductive histories—a meddling interest that echoes into the present. Moreover, these same investigations contributed to assaults on Pitcairn (and now Norfolk) Islanders' cultural and political autonomy, including attacks on their language and restrictions on self-rule.[4] And yet, the purported subjects of "the experiment" were by no means the passive recipients of colonial investigation or mistreatment. The archive of administrative investigation is replete with moments in which the subjects of imperial scrutiny spoke back, on and about the colonial record.

THE EXPERIMENT IS SET IN MOTION

William Denison was considered well placed to carry out the task. A military engineer, he rose through the ranks of the officer corps before assuming the lieutenant governorship of Van Diemen's Land, where he had already managed one British experiment in forced resettlement: the relocation of roughly fifty aboriginal people from Flinders Island to Oyster Bay.[5] Denison was the first figure to consistently deploy the word "experiment" to describe the management and migration of the Pitcairn Islanders; it was following his usage that it became a standard appellation.[6] "Experiment" was a common word in Victorian parlance; it had long acquired its scientific meaning, but it retained a slightly larger metaphoric purchase than it presently enjoys.[7] In keeping with that broader meaning, Denison proposed his experiment principally as a "moral" rather than scientific test in race and colonial settlement. He managed its parameters carefully, however, suggesting that Norfolk Island be reserved for the Pitcairners so that their much-vaunted isolation could continue. "It would in my opinion be advisable," he wrote in 1855, to prohibit any sale of land or right of immigration to anyone but "the race now about to inhabit this small island, and to hold out as little encouragement as possible to the domestication of the other races, be they white or coloured among them, until the effect of the removal of the altered circumstances in which they are about to be placed can be clearly seen."[8] Here, the much-celebrated notion of the Pitcairn Islanders as living at an isolated remove from the wider world was joined to a racialized vision of Pacific settlement, all parsed in the language of controlled experimentation.

In 1855, as ships transported the last of Norfolk's convict population to Van Diemen's Land, Denison dispatched a Royal Navy vessel to Pitcairn Island to conduct a final investigation into its affairs. The *Juno*, under Captain Stephen Fremantle, called on the island in the same manner naval vessels had for almost half a century. Boats set out from Bounty Bay, and then a landing party came ashore and climbed up to Adamstown. Nearly the entire population of 187 gathered to hear Fremantle read a letter from the British consul in the Society Islands, who promised that, though Norfolk could not be formally ceded to the Pitcairners, it would be theirs for settlement. The consul added that his warmest wishes for their prosperity were coupled with "a hope and a belief that they will take with them and retain that sterling principle and noble simplicity which have caused the Pitcairn Islanders to be so cared for by England."[9]

The Pitcairners were unimpressed, and Fremantle noted that they took up the question as "a melancholy subject for deliberation." Indeed, we can read in his reports the record of imperial coercion as well as agentive deliberation and resistance. Many of the older residents recalled the British government's attempt to relocate them to Tahiti in 1831, a move that resulted in seventeen deaths.[10] Others worried that the elderly, the very young, or the sick might not survive the journey. George Hunn Nobbs, the pastor, spoke against leaving. Other outsiders resident on the island, John Buffett and John Evans, evinced the same sentiment but suggested that they would accede so that their children might have a better life. At the end of the night, Fremantle tallied that 153 of the 187 had agreed to the transfer.[11] However, while Fremantle was quick to suggest that an overwhelming majority of the islanders supported the move, a historical reading of his reportage and their decision is necessarily complex. Certainly, some Pitcairners did not want to leave, others may well have felt compelled by the presence of a naval expedition, and still others wanted to pursue the opportunities a move to Norfolk afforded.

After his visit, Fremantle informed Denison of the islanders' decision, adding as a caveat that they were determined to maintain their isolation and independence. He wrote: "The islanders express a hope that they may be allowed to live on Norfolk Island in the same seclusion from the rest of the world as they have hitherto done at Pitcairn; and it is impossible for anybody who has once been an eyewitness of the exemplary conduct and the pious, single-minded character of these innocent people, not to urge a compliance with a request so natural and reasonable."[12] Fremantle's portrayal of the island as pious, loyal, and benevolently isolated was in keeping with established tropes about Pitcairn, historicized in chapter 1. However, though filtered in the moral language and worldview of a British naval captain, this note should also be read as an explicit demand for political autonomy. What Fremantle described as "seclusion" was, by another framing, a desire for freedom from British colonial control. Denison would instead enforce measures designed to maintain the island's "seclusion" rather than political autonomy.

A few months later, on April 22, 1856, the *Morayshire* arrived off Pitcairn. On board was British naval officer G. W. Gregorie, who explained that he would oversee their transfer. Denison had given Gregorie very careful instructions: he was, above all else, to "maintain the strictest separation between the crew of the vessel and the female passengers," as the governor had warned him that the islanders were especially susceptible to "receiving impressions from others." The barrier between passenger and crew was principally meant to serve as a bulwark against sexual liaisons, a measure to maintain what he understood as the "purity" of the Pitcairners in racial and moral terms. Denison also instructed Gregorie to produce a full report of his interactions with the islanders; it was to include any "suggestions which you may think calculated to facilitate the working of the experiment about to be made, or which may tend to the happiness and prosperity of the very interesting people who are the subject of that experiment."[13]

Gregorie's first act was to assemble the community in the Adamstown schoolhouse, where he read a letter from Denison.[14] It warned the islanders that after the *Morayshire* left sight of Pitcairn, the island would fall outside of the sphere of British protection, and men of war would no longer see to the welfare of those who chose to stay. The letter concluded by encouraging the islanders to "maintain that happy simplicity, and that true Christian character, which has gained you the regard and esteem of your fellow countrymen all over the world."[15] The islanders remained skeptical, and many now hesitated. George Adams, one of the older men on the island, had long ago sworn he would never leave Pitcairn. Charles Christian, Jacob Christian, Vernon Young, John Quintal, and their families joined Adams in his skepticism. Gregorie thought their reasons "foolish" but did what he could to persuade the holdouts. Their children, he reiterated, would have a better future.[16] Afterward, Gregorie held a private conversation with the aging George Adams, who confided in the British officer that his grandchild, a young infant, had taken ill and he could not countenance the thought of burying her at sea. If her health rallied, he said, he would join the others on the *Morayshire.*[17]

At six the next morning, the bell in Adamstown rang, and the islanders began preparations for departure. The *Morayshire* had arrived earlier than expected; the Pitcairners had planned to leave in August, and their taro was not yet ready, but they gathered what plants they could, alongside their clothes, cookware, books, and some pieces of furniture, carrying them down the steep, winding trail to the landing.[18] As they watched boatmen shuttle cargo to the waiting ship, Vernon Young and Jacob Christian admitted that they had changed their minds and would go to Norfolk, too. George Adams and Charles Christian remained obstinate in their desire to stay.[19] During the next few days, the Pitcairners continued the arduous work of hauling their belongings, an act made all the more treacherous by rains that turned the steep path into a slippery channel of mud. As the weather worsened, the *Morayshire*'s captain reduced the number of transfers each day and allowed his men to rest on the island each night. Gregorie, enforcing "moral" seclusion, warned the sailors to keep their distance.

On Sunday, the captain himself came to shore, and the island's pastor, Nobbs, preached sermons tailored to the move. As the day of departure neared, only George Adams still swore to remain.[20] The following Tuesday, as the village emptied, he awoke to find that his granddaughter had taken a turn for the better. Only then did he finally acquiesce. Finally, on

May 3, it came time for the islanders themselves to embark. The weather was clear at 9:00 a.m. when the bell at Adamstown rang for what everyone knew to be the last time. Families made their way to the landing, and George Adams's sick grandchild went first into a boat. By late afternoon, the embarkation was done. Finding himself blessed by a sudden surplus of time, Gregorie ordered the *Bounty*'s cannon loaded, too. "I thought the Governor or someone might consider it worth keeping," he wrote; "the islanders themselves do not value it in the least and laughed at me taking such a useless old thing away."[21] The heavy detritus of the *Bounty* myth perhaps meant more to the naval officer than to others. At five o'clock that evening, the *Morayshire* set sail for Norfolk Island, and the looming, green-shrouded volcanic mass of Pitcairn retreated into twilight.

The journey from Pitcairn to Norfolk was an unpleasant one; many of the islanders were seasick.[22] At night, they crowded the lower deck; Jonathan Adams, the father of George Adams's sick granddaughter, complained that the hot and humid conditions there were too dangerous for her, and the captain gave her a cabin, a privilege otherwise only afforded to Nobbs, who emerged from his private quarters each day to read morning and evening prayers.[23] On Sunday, June 1, a baby was born on the ship; its parents named him Reuben Denison, after the lieutenant governor who was the patron of their exodus.

The islanders first sighted Norfolk on June 5. Two naval vessels, the *Herald* and Fremantle's *Juno,* were there to greet them and to assist in their landing. Within days, all of the Pitcairners were on shore. They settled into the prison colony's abandoned houses and barracks.[24] Sailors distributed rations of beef, biscuits, tea, and sugar.[25] Royal Navy officers assigned plots of land and herds of cattle to each family, and a handful of convicts and their guards, who had remained to serve as caretakers, instructed them in the farming and husbanding practices that would form the material basis of their new life.[26] Sometime in those first few weeks on Norfolk, George Adams's granddaughter died. She had fallen ill again shortly after the landing and never recovered.[27]

While Gregorie and Fremantle, per their instructions from Denison, were prolific journal keepers during the transit, there are fewer descriptions of the passage and initial resettlement written by Pitcairners. John Buffett kept a diary during the affair, but his accounts are more matter-of-fact; his writing was in the style of the island's long-standing mode of recordkeeping, which chronicled important events like births, deaths, and ships' visits. His description of the first day on Norfolk reads: "June 8—Landed on Norfolk Island after being thirty seven days on board the Morayshire i.f. London, Captain Mather by whom we were treated with great kindness during the passage one birth no deaths. When we arrived HMS Ship Herald was laying off and surveying, the captain on shore. The Herald's boats assisted us in landing."[28] Several other accounts by islanders survive, such as a small correspondence between Catherine "Kitty" Christian and HMS *Herald*'s second master, Frederick Howard. She described the problems the community faced growing food, as well as a lingering sense of disappointment. "We are not so happy as you may suppose," wrote Christian in October of 1857. "This island is not like our old home."[29]

Fremantle and Gregorie, for their part, both noted in their reports that the mood after landing was subdued. Norfolk was, the islanders said, larger and more open than their old home, without Pitcairn's dramatic precipices and dense vegetation.[30] Nonetheless, Denison interpreted his subordinates' reports as positive and forwarded them to the Colonial Office. He also used the occasion to demand an end to any charitable assistance from outsiders, including the British state. In a letter to the new secretary of state for the colonies in 1856, Denison asserted that "it is better that as little interference as possible should take place with the islanders at present," and he insisted that as soon as the first harvest came in on Norfolk the government cease shipments of any supplies of any kind.[31] The "experiment," as he saw it, was now in motion.

PITCAIRN ISLANDERS AND VICTORIAN SOCIAL THOUGHT

We will come to the colonial state's assessment of Denison's experiment, and Pitcairners' contestation of it, in a moment. First, however, we should examine the intellectual context within and the criteria by which its supposed success or failure would be determined. The hundreds of texts published by sailors and religious writers detailed in chapter 1 had already made the Pitcairn Islanders into a well-known case for writers and intellectuals, and they appeared often in scientific debates about the nature of race and population, usually as a ready-made example or counterexample. By surveying the varied uses made of Pitcairn Islanders by the theorists of the Victorian world, we can open up the intellectual context in which Denison and his successors operated and suggest something of the intellectual stakes in play. Duncan Bell has described this period as "an age of grand (and grandiose) theorizing" in which "it is very hard to separate 'the political' (or 'political theory') from other domains of nineteenth-century thought."[32] In the case of Pitcairn *as* a case, his observation rings true: Pitcairn figured in discussions that, in a characteristically Victorian mode, compressed the moral, the political, and the scientific in sweeping debates over race, reproduction, identity, and governance.

Well before "the experiment," Pitcairn Island was attracting attention from geographers and political economists. Sir John Barrow, the Admiralty figure who wrote the first comprehensive history of the island, was a founding member of the Royal Geographical Society and introduced correspondence about the island at meetings.[33] Intellectuals like Barrow were often especially interested in Pitcairn's rate of population growth, no small concern in the first half of the nineteenth century. In a rapidly industrializing Britain, observers worried that growing populations would outstrip the resources needed to sustain them, and Thomas Malthus's theories accrued a powerful currency—indeed, Malthusian concerns drove both of Britain's forced relocations of Pitcairn Islanders during that period.[34] Malthus himself considered islands useful models for thinking about limitations on space and resources. In his *Essay on the Principle of Population,* he devoted an entire section to the Pacific and noted that with its own finite limits, "the whole earth is in this respect like an island."[35] With the "discovery" and settlement of Pitcairn, theorists could point to a real-world enactment of Malthusian thinking, especially as the island's tremendous rate of population growth seemed to demonstrate perfectly his law of the geometrical increase of human population.

Accordingly, Malthus's critics were compelled to address the Pitcairn case in their arguments against him. Reformist politician Thomas Doubleday took up Pitcairn as a counterexample in his anti-Malthusian tract, *The True Law of Population Shewn to Be Connected with the Food of the People.*[36] He admitted that the Pitcairn Islanders had increased their population dramatically, but suggested their growth was won not in spite but because of the constraints Malthus had warned should lead to famine and death. Doubleday described the Pitcairners as living a hardscrabble existence, arguing that their "constant exertion" made them strong, unlike the spoiled nobility of England who lived unlaboriously and reproduced at a much lower rate. Like many social-theoretical utilizations of the island in the mid-Victorian period, his argument was centered on the islanders' morality; they lived well and so reproduced well. Others agreed. "These colonists were placed in the exact position to try this grand experiment completely," remarked a favorable critic in the *British Quarterly Review.*[37] Political economists, in their discussions of Malthus, contended with the example of Pitcairn for years.[38] Henry George, in his criticism of *The Principle of Population,* admitted that Pitcairn might offer a rare example of pure Malthusian increase but insisted it was "an exceptional case."[39] One of George's own critics, Lord Bramwell, in turn thought it anything but an exception. With its own finite parameters, he declared, *"the world is Pitcairn's Island enlarged."* His assertion is one of the clearest expressions of Pitcairn as microcosm I have yet encountered.[40]

The island's population figures also became very useful in nineteenth-century scientific debates about race. During the middle decades of the nineteenth century, ethnologists took "polygenic" stances, especially in France and the United States, arguing that humanity was broken into distinct races with separate origins, perhaps even constituting different species.[41] Though "species" was a markedly fluid concept, many definitions suggested that its boundaries were discernible in terms of fecundity; the members of two different species, even if they could reproduce together, would not bear fertile offspring. The infamous race scientist Josiah Nott insisted in 1844 that hybrid populations were not viable in the long term. "If a hundred white men and one hundred black women were put together on an island, and cut off from all intercourse with the rest of the world," he wrote, "they would in time become extinct."[42] As in the case of Malthusian theory, the isolated island was considered to be the ideal thought experiment with which to imagine population dynamics at their most absolute. And, as in the case of Malthus, critics of Nott used Pitcairn as its living embodiment. There, wrote a detractor in the *Southern Quarterly Review,* its Anglo-Tahitian population "regularly increased up to 1840," hardly the extinction Nott prophesized.[43]

It was in Britain, where the mythos of Pitcairn was the strongest, that the island appeared most often in responses to assertions about the infecundity of mixed-race populations. Britain had its polygenic school, too; the Anthropological Society of London in particular welcomed polygenic interpretations. However, with a long tradition of liberal and evangelical sympathies, many members of the rival Ethnological Society of London tended toward "monogenism"—a theory proposing a single origin for humanity. The two societies often clashed.[44] In 1864, F. W. Farrah read a paper at a meeting of the Anthropological Society that drew from prominent polygenists such as Louis Agassiz and Paul Broca to argue for the general infecundity of hybrid populations.[45] Alfred Russel Wallace led the monogenist rebuttal, arguing that "there was the well-known case of the Pitcairn Islanders."[46] Another member joined him to affirm that "the evidence of the Pitcairn Islanders contradicted the assertion that the progeny of mixed breeds are infertile."[47] A few months later, John Crawfurd offered a paper at the Ethnological Society in response and again relied on Pitcairn as a ready-made experiment and living counterexample: "A chance not likely to occur again furnishes us with a fact which seems to me to set for ever at rest the allegation of infecundity in human hybrids."[48]

Crawfurd suggested instead a common alternative notion: "hybrid vigor." Rather than tending toward infecundity, its proponents argued, the offspring of "crossed" races were actually healthier. It was a theory that nonetheless still abided notions of European superiority; in such crossings, "lower races" were biologically "uplifted" toward Europeans. It was a pressing concern for some mid-Victorian observers at a moment when the settler empire was fast expanding. As organizations like the Aborigines' Protection Society were keen to indicate, Indigenous peoples were dying in the face of European colonialism, not least Tasmanians under Denison's rule, who were a much publicized case. An 1864 essay read in Oxford put it in clear terms: "Is it a law of nature that the progress of one race should be founded on the extinction of another?" But it offered an alternative: biological and cultural amalgamation; "Pitcairn's Island," added the lecturer, "seems to confirm this solution."[49]

It was a way of thinking that built on Darwinian ideas of "fitness," and Darwin himself had some interest in the islanders. He read Belcher's *Descendants of the Mutineers* alongside Denison's autobiography to determine the Pitcairners' population numbers, using them as an index of their vitality. When Darwin found Denison's published sources insufficient, he wrote to the governor's widow to find out more:

> It is notorious that these people increased rapidly in Pitcairn I., & the point which I am anxious to learn is whether, after their removal to Norfolk I, they continued to increase as rapidly as before. I fear that it is hardly probable that I could any how ascertain their numbers when first brought to Norfolk I, the number of those who left the island, and their number at any recent date. In the animal and vegetable kingdom it is certain that very slight changes of condition sometimes affect the fertility of species in a marvelous manner, and I much desire to ascertain whether any thing of this kind has occurred in Norfolk Island.[50]

Darwin had already used Denison's description of the Pitcairners' migration as an "experiment" in earlier editions of his *The Descent of Man* to argue in favor of hybrid vigor. Contrasting the happy fates of the mixed-race Pitcairners and the extinction of the aboriginal Tasmanians, he wrote: "It appears that a cross with civilised races at once gives to an aboriginal race an immunity from the evil consequences of changed conditions. Thus, the crossed offspring from the Tahitians and English, when settled in Pitcairn Island, increased so rapidly that the island was soon overstocked. . . . What a contrast does this case present with that of the Tasmanians; the Norfolk Islanders increased in only twelve and a half years from 194 to 300; whereas the Tasmanians decreased during fifteen years from 120 to 46."[51]

Alfred Russel Wallace, Darwin's codiscoverer of evolution by means of natural selection, maintained a similar interest in the Pitcairn Islanders, and celebrated the possibilities of the Norfolk Island "experiment." In 1871, when another history of the Pitcairners was published, he penned a favorable review. "It is so rarely that social problems can be subjected to anything like a critical experiment," he wrote, "owing to the impossibility of eliminating the disturbing influence of adjacent populations." But, "in Pitcairn Island we had an instance of people almost completely isolated from the rest of the world, who, owing to such exceptional circumstances as can hardly be expected to occur a second time, were morally and physically healthy, with most of the capacities and virtues, and but few of the vices of civilisation; and who were both able and willing to keep themselves free from intermixture or social contamination. Many curious problems were here in process of solution."[52] Both Wallace's and Darwin's examples demonstrate how clearly the island's supposed moral, racial, and scientific exemplarity were intertwined, as well as the ways governmental intervention had already contributed to their experimentalization in scientific discourse.[53]

However, by the same token, Wallace worried that "the experiment" did not, in fact, make for a terribly good experiment—as it was far from "natural" by any definition. On Pitcairn, he wrote, they had been "one of the happiest, as they are one of the most interesting, of English-speaking communities," but removal to Norfolk Island brought them closer to what he regarded as outside influence. In the interests of science, he urged the government to keep the islanders as isolated as possible—including a call for their continued self-rule. He wrote: "We trust that the officious though well-meaning persons who so rudely broke up the happy and united community in its original home, will for the future leave these interesting people to manage their own affairs (for which they are quite competent), and to work out after their own fashion the many problems in physical, social, and political science which increasing population will soon force upon them."[54] In Wallace's description, and his warnings, decades of social thought came together to understand the Pitcairn Islanders' removal to Norfolk as an experiment in progress, one whose criteria for success or failure would center on economic management, population dynamics, and racialized notions of biological reproduction. However, its ongoing success as well as an assessment of its results were now in the hands of the British colonial administrators. It is to their investigations, to the discourse that accreted around them, and finally to Pitcairn Islanders' responses to them that we now turn.

THE PRACTICES OF COLONIAL INVESTIGATION

In 1857, just a year after the Pitcairners' arrival, William Denison traveled to Norfolk Island to investigate the progress of the "experiment." Like many investigators on Pitcairn and Norfolk before him, he began his visit by addressing an assembly of the people. He took the opportunity to scan the crowd's faces, searching for what he understood as physical signs of their heritage. "I was struck," he wrote later, "[that] there were none who could be called strikingly handsome, but all, or nearly all, were good looking, had good features, well-developed foreheads, and an intelligent expression of face." Summary anthropological judgments like these were common in colonial reports, which reflected the attitudes of racial science then gaining cultural ascendency. Denison's condescending assessment, like those of the colonial administrators who followed him, would compress physiognomy, morality, economics, and much else together into a singular judgment of their success and capacity for colonial self-rule.

Denison's early assessment was mixed. He confirmed the moralizing descriptions of his predecessors, insisting that the islanders' much-vaunted religious purity still formed "the very best foundation upon which the structure of a society can be raised." However, Denison did note with alarm that the islanders were already beginning a gradual decline into what he and his contemporaries termed "indolence." The chauvinist notion that non-European people, especially Pacific Islanders, were less capable of or inclined to work already had a long and harmful history by the 1850s.[55] Colonial intervention was often justified by pointing to supposed Indigenous underuse or misuse of resources, though by the nineteenth century figures like Denison would increasingly reconstrue "indolence" as a racialized trait.

On Norfolk, Denison would deploy that discourse to render profound changes. He urged that the government of New South Wales disrupt the islanders' tradition of communal property; instead, he would further allocate land to individual families on Wakefieldian lines as an inducement to private labor.[56] As the originator of the "experiment" on Norfolk, Denison placed a primacy on maintaining the islanders' isolation, but he was not especially concerned about maintaining traditional ways of life. Rather, he was the first of many governors who would judge the islanders by British understandings of "industry" and "economy," those quintessentially Victorian values that viewed self-sufficiency in moral as well as material terms. Reflecting the broader discourse outlined in the previous section, economic productivity was intertwined with racialized notions of sexual propriety, the chief metric by which mid-Victorian officialdom would assess the "success" of the Norfolk Island experiment.

Denison's reports were, like other government narratives about the Pitcairners, compiled and published. Excerpts from them in turn reappeared in popular accounts in the subsequent decade, especially Lady Diana Jolliffe Belcher's *The Mutineers of the Bounty and Their Descendants in Pitcairn and Norfolk Islands,* which collated much of the existing official correspondence and went through multiple editions. These in turn formed the basis of some scientists' interest in Pitcairn, including Wallace's.[57] From that published material, readers could learn, for instance, that Denison made a return trip to Norfolk Island in 1859 and discovered that the community continued to struggle. Some families chose to return to Pitcairn, bartering for passage on visiting ships, an echo of agentive decisions to disrupt a similar colonial relocation project a decade before. Denison encouraged the remaining Pitcairners to stay on Norfolk, reminding them that they would receive no aid or protection if they returned to Pitcairn.

The following year, Denison left Sydney to become governor of Madras and then governor-general of India. The "experiment" would now be administered by his replacements. The first, Governor John Young, visited in 1861. He met with the islanders and walked Norfolk's hills, inspecting farms and homes. Like Denison, he attempted to force Norfolk's isolation. He suggested, for instance, that the bishop of New Zealand, John Patteson, be dissuaded from opening up a Melanesian Mission school on the island—which he eventually would do in 1867. Arguing for the school's exclusion on the basis of race, Young insisted, "I cannot conceive of anything more likely to demoralize the population and turn it from a high type of race it now assumes back to that of mere South Seas savages. Looking at the simplicity and inexperience of the young girls," he added, "I cannot but think that the introduction of a number of half-savage youths at the period in life when their passions are least under control would be in the highest degree pernicious, and indeed fatal to the prospects of the community."[58] Here, Young made explicit what was often gestured at in Victorian governmental reports: the centering of the anxious colonial gaze on sex and reproduction as the key elements of a racial worldview, one in which "degeneration" for the first time began to loom as an anxious specter. In particular, his comments indicated the ways the "experiment" treated the Pitcairn Islanders' status in imagined racial hierarchies as contingent, threatened by a potential backward "turn" if Pacific influences were not controlled or contained. This was "colonial common sense," a belief that only racial disciplinings of reproduction, expressed as disciplinings of "morality" and sex, could ensure a successful settlement on Norfolk.[59]

Young's successors would add to that discourse, with profoundly adverse consequences for the community's continuing political autonomy. Formal administrative reports grew increasingly derisive across the following decades, and began to deploy racial and eugenically inflected discourse in order to question self-rule. An 1884 report by special commissioner Henry T. Wilkinson, who lived among the Norfolk Islanders for six months, was especially negative, again describing the islanders as "indolent."[60] In 1895, Governor Viscount Hampden surveyed the literature produced by Norfolk's colonial bureaucracy in similar terms: "The general purport of these reports is that the island is fertile, but that the population, partly from inherited character and habits, and partly from the absence of any controlling party, have no desire to cultivate the land."[61] He recommended that New South Wales, as part of a newly federated Australia, take direct control over the island.[62] The Norfolk Islanders vehemently opposed any such plan and wrote a series of petitions to the New South Wales government, the Colonial Office, and the queen in protest. Such a move, they implored, would "involve the destruction of the distinctive character and race of the people, as well as of their ancient laws, institutions and customs of themselves and their Pitcairn fathers. . . . It would inundate the island with strangers."[63]

However, governmental oversight and an attendant eugenic and "moral" scrutiny by administrators only increased. Among the most intense investigations was a 1903 visit by New South Wales official William Houston, whose report was deemed so focused on matters of sex and reproduction that it was withheld from publication. It survives in the archives of the Colonial Office, which contain not only Houston's full report, but also verbatim accounts of every meeting and interview the governor held, about seventy across the space of a month. They constitute a remarkable archive of colonial intelligence gathering. Houston's report demonstrates well the profound scrutiny such visits entailed, as well as agentive local resistance to them. Indeed, Houston wrote of his first public meeting that a palpable "tenseness of feeling existed, and that antagonism was in the air."[64] In his account, we can locate the eugenic logic of colonial rule, but also the political and cultural stakes of investigations like his—and how well these were understood and resisted by those on whom he came to report.

Houston was ostensibly sent to Norfolk to conduct a survey of land use, but those plans shifted dramatically after he received an anonymous letter by an author claiming to have lived on the island for a year. Echoing a long discourse of paranoid anxiety, the letter asserted that the islanders were masters "in the art of lying and hypocrisy. . . . I consider Australia's confidence men and sharps could not hold a candle to these people."[65] Its author reported that the island was supposedly rampant with illicit abortions and illegitimate births, and encouraged Houston to investigate. From that moment on, the deputy governor decided that he would radically broaden his ambit, and that questions of land tenure would be subsumed by more fundamental inquiries into the island's "moral" life. That a government mission could be so radically and quickly reoriented on the basis of the aspersions cast in a single letter is indicative of the power of the discourse surrounding disciplinings of sexual reproduction, both on Pitcairn and Norfolk and in the British Empire more generally.[66] Questions of "morality" and reproduction were at the heart of Denison's "experiment," and took precedence over all else.

However, the demeaning language of duplicity the anonymous letter exudes, and Houston's credulous response to it, should not be read as an archive of artifice. Rather, it is best understood as an artifact of imperial derangement, a paranoid recognition of local agency transposed and expressed in the language and logic of investigation, in idioms of scrutability and obfuscation. Observers like Houston and his informant knew that the community had been visited by outsiders like themselves for almost a century, and from that recognition of iterative colonial meddling they constructed an expectation of subterfuge. Likewise, the focus on abortion and births outside wedlock—nominally unlawful but haphazardly prosecuted across the empire—was representative of a larger turn toward anxieties about and disciplinings of gender, sexuality, and medicine during this period, as well as the social-theoretical discourse emerging about the islands on the part of Victorian armchair intellectuals.[67] It, like all discussions of "morality," should be read not as an archive of "reality," but instead as the constructed product of imperial obsession. Let us examine Houston's practices on Norfolk in detail, then, to see how outsider obsessions resulted in specific forms of colonial scrutiny and interference—as well as to locate Indigenous contestations of the same.

We begin with a scene of panoptic surveillance. A week after his arrival, Houston stood before a map of the settlement alongside the island constable (Collins) and the court magistrate (Nobbs):

> Mr. Houston: (To Collins) Anything you tell me about these people will be confidential, and Mr. Nobbs, you know, is a gentleman who would respect that.

MR. NOBBS: Oh yes, sir, absolutely.

MR. HOUSTON (PRODUCING A PLAN): I have got all the different buildings numbered on this plan.[68]

One by one, the assembled party worked their way through the island's houses, in each instance determining which family owned and lived in it (I have anonymized their names using dashes). However, though a trained surveyor, Houston looked over the chart with an eye not to its physical, but rather its moral topography. They turned to a house.

MR. THOMPSON: [Indicating house number] Occupied by——,[69] and his son-in-law——. Pitcairn Islander. Occupied since transfer from Pitcairn.

MR. HOUSTON: Are they still there?

CONST. COLLINS: Yes,——is still there.——is not there now. That is still his home.

MR. HOUSTON: What character have these people got?

CONST. COLLINS: I always find them very good.

A few houses later, the conversation proceeded:

CONST. COLLINS:——is alright, sir. He has two daughters with illegitimate children, one two and the other one.

MR. HOUSTON: One of the schoolteachers is named——?

CONST. COLLINS: That is one of the illegitimate children.

MR. HOUSTON: She is a very decent little girl, too, apparently?

CONT. COLLINS: Yes.

MR. F. NOBBS: She is a very nice girl, but the family have been very loose in that way.

Houston's secretary made a note of it, and they continued on again.

After working through every household on the island, the deputy governor arrived at a "moral" survey of the entire population, having, through court records, the constable's memory, and abject hearsay, divined some accounting of the community's supposed moral and reproductive landscape. It was a pseudo-empirical assessment, useful to him for addressing a century of debate about the island's celebrated purity and rumored degeneration. It offers a window into the variety of knowledge British officials pursued on the ground and the evidence that comprised their reports, particularly at a moment when the policing of sexual, reproductive, and racial boundaries became a more pronounced imperial project. In particular, it shows just how invasive these investigations could be, as investigators assessed Pitcairn and Norfolk islanders' lives with a granularity and intrusiveness that continues into the present.

Houston by no means restricted his inquiries to the constabulary, however. He conducted interviews with locals of every age and gender, though recent immigrants rather than Pitcairn descendants were most eager to speak. The deputy governor's questions reveal a great deal about his interest in sex and reproduction, as well as his growing paranoia over truth and obfuscation. Take his conversation with one anonymous "outsider," anonymized by his secretary as Mr. 6:

> MR. HOUSTON: You say that the moral condition of the island is what?
>
> MR. 6: Well, it is really bad . . .
>
> MR. HOUSTON: Where could I get any direct evidence of these things? Where could I go to enquire about them? I intend to enquire of 7 and 8 and 9: but how could I get a knowledge of the facts? Would it be possible? I doubt whether it would be possible.
>
> MR. 6: As I say, I have been in the houses: I have seen people go into the room and shut the door; and I have known what has been going on; but I cannot prove it. You cannot absolutely prove it. I have seen things when riding in the bush. I have seen young fellows and young girls get off the ground, but I cannot prove anything, can I? You see they are cunning as can be. Of course the names are known of people here that carry on these things.

Houston's transcripts are laden with such artifacts of colonial and patriarchal anxiety, converting rumor into an imperial archive of racializing and moralizing reportage. Again, we should read these breathless accounts as the result of the existing discourse surrounding the islands. Regarded by writers and administrators as religiously or morally separate from the world, but also as threatened by racial "admixture" or "degeneration," they increasingly became subject to governmental scrutiny—especially when it came to the assessment of Denison's "experiment."

Houston, for instance, increasingly became convinced that the islanders' heritage was the direct cause of their present "problems." He sought out signs of recessive "Tahitianness" lurking in the shadows of the island culture. In one interview, he asked about rumors recorded in previous government reports that the islanders maintained heterodox burial practices, which he thought informed by their Tahitian ancestry.[70] In another interview, Houston discussed the islanders' so-called animal propensities, an especially demeaning Victorian euphemism for what he considered sexual impropriety, and whether they could be attributed to Pacific ancestry, intermarriage, or both.

Mr. Houston: As the result of Intermarriage is there not usually a declension of the mental powers?

Mr. 55: I think so.

Mr. Houston: And is there not a special physical development?

Mr. 55: Oh yes.

Mr. Houston: Would that not account for some of the animal propensities?

Mr. 55: I should be inclined to think that the animal propensities result from their descending from the Tahitians, but there is mental and physical degeneration very marked.

Mr. Houston: Physical degeneration?

Mr. 55: Yes, such things as the increase of Tuberculosis and diseases of that kind, and the increase of mental diseases. For instance, there is a child born that will be a blind idiot. And in one family, No 58, there is one son who has tuberculosis spine with curvature and distortion, and another whose mental capacity is not by any means what it ought to be, and a child who is virtually a blind idiot . . .

Mr. Houston: That is a family where there has been close intermarriage? . . . It could be traced?[71]

Houston sought what information he could on the islanders' physical health; he procured from informant number 59 a list of the islanders' "deformities, etc."[72] Together, these concerns over the body and degeneration, understood as the product of heritable traits passed down from a latent, feminine, "Tahitian" influence, presaged the investigations of professional physical anthropologists and eugenicists in the coming decades. In particular, the notion that racialized traits could be legibly traced would remain a hallmark of investigations for decades to come.

After a month of investigation, Houston wrote a report summarizing his findings, while his secretary compiled the minutes of each interview. His conclusions were highly racialized. "Their immorality or want of morality," he wrote, "is to a large measure due to the Tahitian blood in their original stock."[73] Many Colonial Office staffers expressed shock at the suggestion that the Pitcairners' carefully preserved utopia had turned to dystopia. A note on the dossier exclaimed: "I wish the details were known—they would put an end to these figments of Arcadian communities," not least because it would provide the government an excuse to deport the population of Pitcairn if needed.[74] Though never published, Houston's report was referenced by subsequent official accounts of Norfolk Island life, most explicitly in 1910 visit by the governor of New South Wales, Lord Chelmsford. Assessing a half century of governmental investigation into the lives of the Pitcairn Islanders and the success of the Norfolk "experiment," Chelmsford summed up the project in no uncertain terms:

> The attempt was made in 1856 to establish [on] an inaccessible island a community free from all the evils of civilization, and to reproduce in the 19th century a Garden of Eden without the serpent. No doubt those responsible for the experiment launched it in good faith, believing that they had found in the Pitcairn Islanders a people endowed with all the virtues and all the innocence of our first parents before the fall. I do not hesitate to say that no experiment has ever failed so dismally, and that the Norfolk Islanders of Today, so far from being innocent and virtuous, are as debased, as idle, as immoral or—unmoral perhaps would be a more suitable epithet—a people as exists on the face of the earth.[75]

Lord Chelmsford had offered, in his mind, the last word. After a half century, Denison's "experiment" had proven, by the moralizing and increasingly eugenic logic of British imperial governance, a "failure." However, rather than treat Chelmsford's assessment as a true accounting of Norfolk Island life, we should instead understand it as the product of a century of British obsessions and a testament to the shifting nature of colonial rule and ambition in the Pacific. Administrators, building on the utopian mythos constructed around Pitcairners, had deployed the machinery of the imperial state in a deranged attempt to manifest their visions. Chelmsford's superfluous and hyperbolic language, infused with the religious metaphor of Eden corrupted, was above all a product of colonial disappointment, a realization that the subjects of Denison's racial "experiment" had not conformed to their fantasies. The result was an inversion of utopian projections. Rather than conceive of Norfolk Islanders as a complex community with political interests, part of and connected to the wider world, Chelmsford could only see the island as paradise corrupted—not utopia, but dystopia. It was a competing image of Pitcairn or Norfolk Island as microcosm that would persist, with haunting consequences for the communities themselves, in the discourse produced by outsiders well into the twenty-first century.

COLONIAL KNOWLEDGE AND ITS CONTESTATIONS

It is not my intention here to dredge up the artifacts of an intrusive investigatory discourse, especially not one animated by racist, eugenic, and imperial concerns, in order to rejuvenate aspersions against Pitcairn and Norfolk Islanders. Instead, we should read the colonial archive to understand how administrative investigations were situated within and abetted structures of power, and the ways their conclusions revealed something about colonial officialdom's changing visions of race, governance, and the Pacific. Figures like Denison and Houston sought to determine the "fundamental truth" about the character of the Pitcairn Islanders not only, or indeed not principally, to rule Pitcairn or Norfolk alone. Rather, they sought to understand how "race mixture" and colonial settlement would play out in what they regarded as a microcosm with lessons for the whole of the empire, within and beyond the Pacific. In so doing, they reframed a vision of Pitcairn Islanders as the inhabitants of an accidental Eden into a vision of Pitcairn Islanders as the subjects of a racialized experiment. Across the next century, a series of academic investigators would take them and their descendants up as research subjects, arguing that their communities constituted "natural" laboratories—and in doing so they would elide the very artificial efforts of figures like Denison and Houston to construct Pitcairn Islanders as experimental subjects in the first place.

Governors like Chelmsford were quick to pronounce summary judgments of their subjects. But what summary judgment should we make of administrators like him? In their reports, obvious threads of power and control are visible, but so are colonial anxiety and unease. The underlying metaphor of the "experiment" that drove their assessments hinged on fantasies of control, isolation, and legibility. On Norfolk Island, investigations like Houston's were enactments of tremendous colonial power, resulting in severe abrogations of promised political and cultural autonomy. However, colonial power also met its limits. Denison had not produced the scrutable, racialized subjects he desired. In seeking to re-refound an imagined Eden, the colonial state attempted to de-range Pitcairn Island, to drag utopia into the worldly. Facing the collapse of fantastical project, figures like Houston and Chelmsford instead turned to dystopia and duplicity as overriding metaphors for making sense of their own frustrations. A fear of practiced subterfuge, in particular, was in its way a recognition of both Norfolk Islanders' agency and the consequences of long-standing iterative investigation by their predecessors. However, that discourse, itself also a deranged product of colonial fantasy, obscured very real moments of agentive contestation of imperial investigation.

That administrators like Houston had power is obvious. In declaring the "experiment" a failure, the colonial bureaucracy came to adopt a distinctly high-imperial vision of race, one that sought explicitly to control the sexual and moral lives of its racialized subjects as a matter of good "science" and good administration. Rigid and biologized conceptions of Britishness, which privileged race and the body, overtook previous markers, not least religion and language. That shift was part of a broader transformation in the way the British imagined their empire and the subjects residing within it, especially in the antipodean colonies. As Warwick Anderson has written, "during much of the nineteenth century, being 'white' in the Australian colonies usually meant claiming British ancestry," but in the last decades of the nineteenth century, whiteness was re-created on the frontier by scientific and medical examinations of the body, investigations that constituted a "discourse on settlement" in their own right.[76] Moreover, in the case of Norfolk Island, race and sex combined as fetish objects of imperial fascination with more stark clarity than perhaps in most other spaces. Literary scholar Anne McClintock has noted that, at precisely this period, "sexual purity emerged as a controlling metaphor," and perhaps *the* controlling metaphor, for "racial, economic, and political power."[77] Sex neatly combined the moral, the racial, and the bodily into a set of signs legible to Victorian investigators. Moreover, because so much of the fascination with Pitcairn Islanders on the part of outsiders stemmed from their famous descent from a founding group of mutineers, genealogical notions of "heritage" and disciplinings of reproduction became particularly dominant not only as subjects of discourse but as objects of colonial investigative practice.

Administrative scrutiny contributed to a colonial worldview that would see the empire increasingly attempt to discipline, in racial and sexual terms, the populations under its purview. Norfolk Island itself would suffer a profound loss of political and cultural autonomy. Chelmsford advised the Colonial Office and the Australian authorities of "the necessity of a paternal government for some time. . . . The islanders," he said, were "so intermarried, that honest, healthy self-government is impossible." He advised that Australians settle Norfolk Island in large numbers in order to swamp the Pitcairn population and thus diffuse its supposed racial and eugenic risks, as well as their political power. It was a move that not only would put a distinct end to the island's isolation and the experiment's parameters, but that was meant to forever dissolve the Pitcairners as a distinct people.[78] At the same time, the Colonial Office quietly instructed ships' surgeons to investigate the health of Pitcairn Islanders in explicitly eugenic terms.[79]

And yet, beneath the stern pronouncements and official reports lurked a profound anxiety that knowledge about Pitcairn Islanders was not only difficult to obtain but deliberately obfuscated, and that the studied were excellent students of their studiers. One informant, in a meeting with Houston, warned him in paranoid tones that he, too, would be watched on the island. They "will come to the verandah without their shoes and listen through the windows: you do not know they are there. They sneak up and you are quite unaware of their presence—one has to be very circumspect in one's speech."[80] Chelmsford warned of the same, and cited the following extract from Houston's report to suggest a purposely hidden, dystopian Pitcairn lurking beneath the surface:

> MR. HOUSTON: But at Pitcairn the strain was stronger than it is here now, and yet everybody that went there depicted them as an isle of saints . . . pious, innocent, simple. How do you account for that?
>
> MR. 50: I have the older people talk here—they are gone now, dead, and I do not think they were anything of the kind. Did it ever strike you what old George [*sic*] Adams did with all those women?
>
> MR. HOUSTON: I never thought about it; but it did strike me that he was left with all those women, and that there was some population.
>
> MR. 50: There were some queer doings at Pitcairn. You see there was a fine to let it be known to the world.

This was a profound misrepresentation of Pitcairn genealogy, which even then was widely known from texts that republished data from the *Island Register.* Even the name of John Adams, one of the most famous of the mutineers, was misremembered. However, this misreading and speculation is suggestive of the ways outsiders projected and interwove concerns about "morality" and "duplicity." Indeed, reports as far back as Henry Wilkinson's 1884 investigation suggested that the Pitcairners colluded to prevent the revelation of unsavory aspects of island life to their interlocutors. According to Wilkinson, "a heavy penalty was inflicted on any one of their number who exposed to strangers facts which might be used to their disparagement."[81]

In colonial discussions on purported subterfuge, we can also locate a certain kind of transfigured recognition that iterative colonial scrutiny left deleterious scars. One particularly stark example of this half recognition is visible in the 1913 correspondence between British consul to Tahiti H. A. Richards, under whose jurisdiction Pitcairn fell, and James Lyle Young, a trader and founding member of the Polynesian Society.[82] Young wrote a scathing account of Pitcairn, which Richards passed on to the Colonial Office. He described the island as a place where devout piousness coexisted "with the utmost laxity in matters sexual." However, he qualified the "hypocrisy" that so outraged his peers. "We must take into consideration," he wrote,

> the fact that for a hundred years past the Community has lived on charity from passing vessels, and the Islanders have long ago learned that the romance surrounding their history, and their reputation for loyalty to the throne, as well as for simplicity and innocence and devout religious

> belief, have always been their assets; their stock-in-trade, in short, which leads to the good opinion of visitors and to consequent gifts. . . . Considering therefore, the whole history of these people, and their descent and conditions of life it is, perhaps, unjust to stigmatize as hypocrisy and deceit, the smooth face, the ingratiating manner, and the profession of religion which these poor people are, so to speak, compelled to put on in order to live.[83]

Young was not a sympathetic friend of the island, and his perception of the islanders was steeped in the anxious language of racial and sexual panic. However, he did identify a fundamental fact of life for Pitcairn and Norfolk Islanders: they were the subjects of periodic and persistent scrutiny by outsiders. That intervention itself had destroyed, he suggested, the Edenic microcosm that attracted investigators in the first place. And yet, he could express that history of investigation, and his recognition of insular agency in the face of it, only as a kind of practiced subterfuge.

The results of the interwoven discourses of legibility and illegibility, of openness and artifice, on Pitcairn and now Norfolk were paradoxes that reveal the power and limits of colonial knowledge and colonial rule. Even after a half century of investigations, and after the deliberate construction of Norfolk Island as a controlled experiment, colonial administrators imagined profound and perhaps undetectable limits to their capacity to access any truths about Pitcairn Islanders on either side of the Pacific. After five decades of conscious experimentation, after persistent surveillance, and after thousands of pages of reports, the islands' governors found themselves administering places that, as far as they were concerned, fundamentally defied their attempts to build knowledge. Indeed, it is striking that a community so deliberately experimentalized should give rise so frequently to epistemic anxieties. And yet, in summing up these efforts, a frustrated Lord Chelmsford declared in 1910 that "there is no place where it is more difficult to find out the real truth than Norfolk Island."[84] A colonial experiment meant to produce racial knowledge had become, for him, a site of derangement.

We should not, however, give Chelmsford the last word. That, by right, should go to the Norfolk Islanders, who were subject to invasive scrutiny again and again. Denison, like Barrow before him, sought to stage Pitcairn Islanders as the silent and compliant subjects of racial experiment in colonial relocation and settlement. Houston and Chelmsford, in turn, portrayed their children and grandchildren as the duplicitous heirs of a mutinous heritage. But buried in their archives are the speaking voices of the investigated themselves. The records of colonial investigation offer numerous instances in which Norfolk Islanders directly countered the governor's assessments. These are not "hidden transcripts," secret or subtle arts of resistance on the part of subaltern subjects.[85] They are open contestations, on and about the imperial record.

Take, for instance, a moment from the Houston transcripts in which he asked several unnamed Norfolk Islanders about their supposed "physical deterioration." His interlocutors were initially confused that the administrator would even field a question on such a topic—they were expecting a conversation about land tenure, the stated purpose of the visit. "That is settled," replied Houston. Instead, he asked about inbreeding:

> There is no doubt that the breeding in and in is doing harm, and that what you really want is fresh blood. There is no doubt about that. We see it in England: we see it in every country where cousins marry: that gradually, not always in the next generation, but in succeeding generations, there is degeneration—all kinds of sicknesses, all kinds of diseases and decrepit people, shaky people, come to the fore. I mean that you have got merely to read history, you have merely to read medical books, and you will find that it is one of the laws of nature you cannot get out of.[86]

One informant responded: "Well, your excellency, facts are stubborn things, and we have ocular demonstration amongst us. I am not an advocate of that (intermarriage) but away from Norfolk Island their children are little puny miserable things, while you will find many stalwart fellows amongst the Pitcairn Islanders."[87] It was an argument that engaged with eugenic aspersions directly, by pointing to local health and vitality—and it presaged arguments made by anthropologists only decades later.

Some islanders used their conversations with the deputy governor to conduct their own interrogation, defending their island against negative portrayals in print. They had read, with alarm, some of the reports written about them in the Australian and British press. One interviewee brought with him a text published in 1896 and read it out loud to Houston:

> MR. YOUNG: [Reading from a text] "The very fertility of their soil is against them. They can live without industry and so they reap not the blessing of toil. . . . The lazy ones are fed by the workers and they know it and continue to be lazy. Cut off from contact with the outside world they have no interest apart from their own, no patriotism, no wish to make a name in the world for their island, or for themselves and their children. Given plenty to eat, enough sleep, and free scope to their lower nature to enjoy itself—they are quite happy and contented. They live in a half lotus-land and they die as they live."
>
> MR. HOUSTON: That is a thing that ought never to have been published.
>
> MR. YOUNG: The Government must know where it came from.
>
> MR. HOUSTON: The Government do not know where it came from.

The government not only knew where it had come from, but it had produced it. The text itself was drawn from reports by colonial officials Hampden, Caruthers, and Oliver—all of whom emphasized the Pitcairners' supposedly environmentally induced "indolence." It was then reproduced in a newspaper, which wound its way to Norfolk Island.

It was another example of the compounding ways investigations and their products accumulated on Pitcairn and Norfolk Islands, and of how understandings of them had shifted. After a century of writing by sailors, administrators, and others, the islands had become famous as purported microcosms of the wider world, places meant to offer utopian—and now dystopian—lessons. The efforts of Denison and the administrators who followed him transformed the myth of utopia into the myth of experiment. Well before any professional scientist made either island into a field site and "natural laboratory," British and Australian governors experimentalized it, inaugurating a century of even more intrusive investigations. The remainder of this book will follow investigations by academic researchers, especially anthropologists, historians, and linguists, into the lives of Pitcairn Islanders, whose studies would shape the development of scholarly fields in the Pacific and beyond. But these researchers inherited not only the myth of the natural laboratory from colonial governance, but also a haunting anxiety and ambivalence over the ability to access the "real" Pitcairn or Norfolk. And, if read against the history of colonial investigation, their fieldwork in a meaningful sense constituted only a few more episodes of scrutiny in long history visits, the material and discursive contours of which were shaped by colonial administrative practices and agentive contestations of them.

CHAPTER THREE

Taking Measure

Anthropologists and the Rearticulation of Race

> Little did the nine mutineers of the HMS Bounty who landed in 1789 on Pitcairn Island and married Polynesian women suspect that they would become human guinea pigs for anthropologists and students of heredity. Cut off from the world, that little colony, which now numbers 200, has made it possible to substitute solid facts for the fancies that underlie racial prejudice.[1]

So began a 1935 *New York Times* article titled "A Racial Laboratory." It offered a remarkable vision of Pitcairn Island as a place where an accident of history had transformed humans into experimental subjects, and where the fallacies of racial prejudice might be easily demystified. It was a vision that contrasted dramatically with that of British and Australian colonial administrators, who had very deliberately and not accidentally constructed Norfolk Island as a racialized colonial "experiment," and who had judged both islands as supposedly eugenically degenerated during the previous decades. And yet, in only a few short decades, the Pitcairn Islanders had gone, at least in the Anglophone imagination, from harbingers of racial doom to beacons of hope.

The author of these changes—and the subject of the *New York Times'* glowing reporting—was the American anthropologist Harry L. Shapiro. Like many of Pitcairn's visitors before him, Shapiro conceived of both Pitcairn and Norfolk Islands as obvious "racial laboratories."[2] Unlike his predecessors, however, Shapiro found in Pitcairn Islanders a set of human subjects who might counter the racial typologies that were then central to anthropology as a discipline. In separate studies of Norfolk Island and Pitcairn Island across the 1920s and 1930s, he would use the techniques of physical anthropology to measure the boundaries between what he understood as English and Tahitian "races," and ultimately to question assumptions about "race mixture" that pervaded not only anthropology but European and American society.

The ways in which Shapiro used Pitcairn Islanders to critique the race science of his moment were complex, embedded in an interwar politics and culture in which the meanings of both race and the Pacific were shifting. By the early twentieth century, growing US imperialism had recentered the Pacific as a field of geopolitical ambition, and once "peripheral" strings of islands now seemed suddenly relevant to readers of the *New York Times,* if still distant and exotic. Pitcairn was a renewed site of interest, too, as series of novels by Charles Nordhoff and James Norman Hall and a Hollywood film starring Clark Gable and Charles Laughton gave the *Bounty* story a fresh prominence.[3] At the same time, the science of race was accruing significance. The American republic and the British Empire saw movements for racial equality and anticolonialism, while far-right regimes across Europe adopted racialist dogma as state ideology. Within science itself, the emergence of genetics had begun to shift the ways anthropologists and biologists studied race, and threatened old ways of understanding human difference centered on typology and comparative anatomy; if individuals carried a diverse host of heritable traits, then "races" were not fixed types at all, but at best nebulous clusters of genes in populations.[4] Against those complex external and internal contexts, physical anthropology used Pitcairn Island, and other field sites like it, to understand the biological mechanisms of racial inheritance.

To make sense of Pitcairn and Norfolk Islands' shifting position in the early twentieth-century scientific imagination, this chapter follows Harry Shapiro's use of the Pitcairn Islanders, tracking his work across several decades and three different islands. The first is Norfolk. Shapiro spent five months there as a doctoral student in physical anthropology, transforming a site of eugenic colonial administration into an anthropometric "laboratory." The next is Pitcairn, where an older Harry Shapiro visited for eleven days in 1934 during an American Museum of Natural History expedition, which would unsettle his scientific ideas about race. The last is Manhattan, where Shapiro worked at the Museum of Natural History, and from which he introduced the Pitcairn Islanders into a wide set of scientific discourses, including anthropology's own insular worlds.

And yet, even if Shapiro and the *New York Times* similarly construed Pacific islands as places of exception where race might be easily demystified, attention to the anthropologist's experience reveals that frictions between the metaphor of the laboratory and the reality of the field gave rise to unanticipated epistemic derangements. Shapiro's physical anthropology was entangled in a history of previous encounters and discourses; long before he ever set foot in Kingston or Adamstown, he dreamed of measuring Pitcairn Islanders, having absorbed the sailors' accounts and the eugenic discourse described in chapters 1 and 2. Moreover, on both Norfolk and Pitcairn Islands, Shapiro found hospitable hosts practiced in the art of welcoming strangers; by a certain accounting, his was simply another visit by another tourist, and we would do well to think of anthropological fieldwork as a set of practices embedded in a larger history of tourism and hospitality.[5] Shapiro's visit and methodologies were also contiguous with a darker and more insidious history of eugenic scrutiny by colonial administrators, who sought to inspect and critique Norfolk Islanders' supposed racial fitness. Unpacking the field practices of anthropology in both places—a history of, among other things, calipers, gossip, measurements, anxiety, and trust—shows that, far from scrutable laboratories that would make plain the mechanics of race, Pitcairn and Norfolk Islands instead begat methodological anxieties. Shapiro found some knowledge easy to glean, some knowledge hard to acquire, and some knowledge, in his mind, inherently suspect or impossible to confirm. Like Chelmsford and Houston before him, the American anthropologist recognized a long history of intrusive investigation and colonial intervention, and it fueled an epistemic paranoia, a worry that nothing could be trusted as true.

Those anxieties in turn shaped the anthropologist's vision of a new and reformed science of human difference. Informed by his experience on Pitcairn Island, Shapiro helped lead a reframing of the race concept in Anglo-American anthropology, fighting against Nazi race science before the Second World War and helping to draft UNESCO's statements on race after it. However, while histories of the decline of scientific racism have usually offered triumphalist accounts in which scientists like Shapiro dismantled "race,"[6] in this chapter I join historians who have begun to problematize that narrative.[7] Tracy Teslow and Warwick Anderson, in particular, have examined Shapiro's work in order to show the ways anthropology shifted away from typological notions of race and toward statistical, population-centered theories of human difference during the interwar years, built in part on a "vision of oceanic racelessness."[8] I do not disagree, though I suggest that it was the frictive disjuncture between Shapiro's visions of experimental legibility and his practical experience with (and deranging visions of) communities suffering long-standing colonial scrutiny that ultimately shaped his rearticulations of the race concept. In the years after his return from Pitcairn, Shapiro pushed back against hard, typological notions of race, but did not do so without reservations. Pitcairn Islanders' deployment as a racial and eugenic datum, even in Shapiro's antiracist work, remained ambivalent, multivalent, and mutable.

FROM GOVERNMENTAL TO ANTHROPOLOGICAL "EXPERIMENT"

As Shapiro told it, he first encountered Pitcairn Island on a winter morning in 1922, as a young student inside Harvard's Peabody Museum. He was born in Boston to poor Jewish immigrants in 1902, and his study of classical languages and classical cello had won him a place at the Boston Latin School and then Harvard. Now, surrounded by the Peabody's collections, he was in his academic element, listening to anthropologist Earnest Hooton deliver several lectures on "race mixture." Amid conjured images of the Pacific, far removed from the cold Massachusetts winter, "with all the magic of the movies the scene shifted, and we were in the warm, vibrant Pacific, on the Cytherean shores of Tahiti, amidst the turmoil of a mutiny on board the Bounty and stranded on a forgotten speck of land called Pitcairn, where human folly was succeeded by inhuman virtue. That unique narrative of an eighteenth-century breadfruit expedition resolving itself into mutiny, court-martial, hanging, crime, murder, and finally, a new population of mixed bloods made a glorious text."[9] Through a constellation of exotic tropes and much-rehearsed myths, Hooton explained that the island represented a near-perfect case study; he told his students that he would "rather go to Pitcairn than anywhere else in the world." Shapiro decided the same and made the island the subject of a PhD dissertation in physical anthropology under Hooton, the first in that field in the United States.[10]

Earnest Hooton, an understudied figure in the history of American anthropology, was enormously influential in the discipline's early years.[11] Historians of anthropology have written a great deal about Franz Boas's legacy as a teacher whose students came to exert a near-total dominance in American social and cultural anthropology.[12] Hooton's corresponding influence among physical anthropologists has been occluded, but it too was profound.[13] His students played an outsized role in the scientific debate over the race concept and fundamentally reconfigured the discipline during the middle decades of the twentieth century. Hooton was in many ways Boas's mirror image. The latter had famously argued that even notoriously racialized anatomical features, like the shape of one's head, were not fixed but mutable categories, and he and his students generally turned anthropology away from biological race and toward cultural relativism.[14] Hooton, on the other hand, regarded race as a semipermeable but nonetheless essential category. The world was, according to him, comprised mostly of "secondary races," derived from a mixture between original "primary" races.[15] He suggested that an investigation of how these secondary groups had evolved, both physically and culturally, would reveal a great deal about the parameters of human diversity and the mechanics of race itself.

For an anthropologist and eugenicist like Hooton, the Pitcairn Islander communities, founded by two separate racial "types," were a microcosm of racial dynamics that had occurred the world over. A close study of their bodies could reveal the underlying mechanics and results of racial hybridization. Perhaps the most important of these was a basic question thought central to the social and political world of the interwar period: Did racial hybridity lead to negative or positive outcomes? Hooton was hardly the first scientist to ask such a question, or to use Pitcairn Islanders to answer it. Charles Darwin saw Governor Denison's experimental, coerced migration of Pitcairn Islanders to Norfolk as a perfect test case, and used his population figures to suggest racial mixture was beneficial. Sir Arthur Keith, Britain's leading physical anthropologist in the early twentieth century, suggested the opposite, especially after examining the living bodies of Charles and Edwin Young, two Pitcairn Islanders who visited Britain with the archaeologist Kathrine Routledge.[16] The question remained open in the 1920s.

To answer it, anthropologists constructed into racial laboratories a number of sites already long subject to intense colonial scrutiny. Germany's Eugen Fischer, who would go on to become a leading architect of Nazi race science, first made a name for himself studying mixed-race communities in what is today Namibia.[17] At the same time, the chief architect of American eugenics, Charles Davenport, worked with Florence Danielson and Morris Steggerda to conduct similar research in Jamaica.[18] Both Fischer and Davenport ultimately decided that race mixture was deleterious, and they used their findings to push for eugenic and apartheid policies in Germany and the United States. "Miscegenation commonly spells disharmony," Davenport declared. "A hybridized people are a badly put together people and a dissatisfied, restless, ineffective people."[19] The stakes were high; measurement of colonized bodies in Jamaica, Namibia, or Pitcairn contributed not only to ongoing colonial violence, but also to the sterilization of bodies in Europe and North America.

Shapiro was motivated by the same concerns and had similar ambitions. However, before going on to study race mixture on Pitcairn, he would have to learn more about genetics. He spent the 1922–23 academic year at Harvard's Bussey Institute for Applied Biology, where he studied with William Castle and Edward East. Castle's model organisms of choice were mice and fruit flies, but he was deeply interested in questions of human heredity; like many geneticists, he worked with Davenport's

Eugenics Record Office. Even before he taught Shapiro, Castle identified Pitcairn and Norfolk Islands as possible living laboratories. "About the only conditions under which a racial cross . . . could be fairly tested would be those under which Pitcairn Island was populated," he wrote in a 1921 textbook, adding, "the experiment has gone far beyond the F_1 generation and would afford unique material for a study of the effects of race-crosses uncomplicated by race-antipathies. . . . It is to be hoped that some student of eugenics will give the case careful and critical study."[20] Shapiro was that student, and Castle, East, and their model organisms trained him in the laboratory work and methodology of the new science of heredity.

Shapiro finally left for the Pacific on July 14, 1923, departing New York by steamer. At the Panama Canal, he boarded the *Paparoa,* a New Zealand Shipping Company vessel bound for Auckland.[21] It took a week to traverse the vast stretch between the Canal Zone and Pitcairn Island. As he approached his destination, Shapiro grew worried. He did not have permission to land, and, without proper papers, the *Paparoa*'s captain was disinclined to let him. As Shapiro later told it, he was equally disinclined to bow to the authority of the captain. In a personal mythology that mirrored Fletcher Christian's, Shapiro planned a mutiny; he claimed to have secretly packed his belongings and conspired with a fellow passenger to "surreptitiously" launch one of the steamer's boats. According to Shapiro, a sudden tropical storm arose, making impossible his plan to jump ship. Shapiro never so much as caught a glimpse of the island.

After arriving in Auckland, Shapiro reformulated his plans. Short on funds and time, he decided to investigate Norfolk Island instead. Shapiro took a steamer to Sydney, and after a month of waiting, finally departed for Norfolk Island. He landed there like most visitors did, at the jetty in Kingston, amid the old penal settlement—were he immediately found a set of hospitable welcomes. On learning that Shapiro was from Boston, older islanders, who fondly recollected American whalers from New Bedford, asked after long-remembered acquaintances. His most fruitful early meeting was with Cornelius "Cornish" Quintal.[22] Cornish was born on Pitcairn Island, the grandson of mutineer Matthew Quintal. By the time of Shapiro's visit, he was one of the oldest men on the island, a patriarch who had spent his youth harpooning whales and his old age leading church services.[23] Cornish served as Shapiro's point of entry to the Pitcairn community, a role he had played for other visitors in the past.[24] Shapiro considered him his "guide and mentor."[25]

A guide was helpful, especially given the island's complicated political and cultural terrain. As was true for all researchers, Shapiro's success depended on the hospitality of his hosts. He had arrived on an island with a complex history. It was certainly not as isolated from outside influences as the "laboratory" metaphor suggested; the Anglican Melanesian Mission had left the island a few years earlier, but "strangers" were hardly unknown. Regular visits by ships brought ever more frequent visitors and even tourists, some 238 in 1923.[26] Many of these were passengers in transit from the New Hebrides to Sydney on the *Makambo* who rode hired horses or hunted game with guides like Cornish; Shapiro was, in a meaningful sense, just another tourist like them, shepherded across Norfolk's landscape. There was, too, a long legacy of friction with the Australian mainland. When Pitcairn Islanders came to Norfolk Island in the nineteenth century, they did so with the understanding that the island would be reserved exclusively for their community. However, after declaring an end to Denison's "experiment," Australia had encouraged outside settlement and increasingly interfered in the island's affairs, which it now nominally governed as external territory.

In consequence, when Harry Shapiro arrived with calipers and questions about heredity, Norfolk Islanders were quite accustomed to outsiders interested in their reproductive and "moral" lives—government agents, working to absorb the island into the Australian state, had already attempted to construe their bodies as eugenically deteriorated. That iterative history of colonial scrutiny derailed Shapiro's plans almost from the outset. He complained that "the conditions of field work made the collection of family data very difficult. I was able to obtain only a few complete family records, so that a study of family inheritance was impossible."[27] Shapiro left behind few field notes from Norfolk—his expedition diary in the American Museum of Natural History (AMNH) archives contains only two entries, one of which outlines the procedure for summarizing a "case in mental disease" and instructions for administering an intelligence test.[28] Above all, Shapiro evinced a paranoid worry that the island's long history of colonial scrutiny had made it very un-laboratory-like indeed, and he expressed that recognition through accusations of subterfuge. Shapiro wrote that the Norfolk Islanders "form affectionate attachments to people of whom they are fond, but I have the feeling that their emotions, although easily exhibited, are superficial."[29] This observation was, Shapiro knew, in keeping with a long-running discourse about secrecy. He noted that "many experienced observers maintain that in spite of the apparent openness of manner of the Norfolk Islanders, it is impossible to gain their entire confidence."[30]

Shapiro's work was intrusive at a corporeal level. Anthropometry, the measurement of human beings, was a simultaneously detached and intimate affair, already fairly standardized by 1922. Shapiro followed the method of German anthropologist Rudolf Martin.[31] He used an anthropometer, which consisted of a rule along which slid two long metal arms, to measure the height of the skull, the width of the shoulder blade, the thickness and breadth of the hand, the length of the middle finger, the thickness of the leg, the width of the foot, the width and depth of the chest, and the span of the pelvis. Using two different calipers, he took detailed measurements of the head and face, including length and width of the skull, several measurements of the nose, and measurements of the jaw. Using a set of cards ranging across an array of human skin tones, he recorded the shade of the islanders' skin, both on the cheek and on the inner arm. He felt their skin, too, to record its texture. He marked down the islanders' hair color, and ran his fingers through it to feel its thickness. He noted their eye color and the shape of the eye folds. He asked the islanders to open their mouths and made records of their teeth.[32] In total, Harry Shapiro took these measurements from 150 people: 64 men, 49 women, and 37 children.

Five months later, Harry Shapiro packed away his calipers. He left Kingston with a precious gift from his subjects, a "poi pounder" said to have belonged to one of their ancestors.[33] Back in New England, he defended his dissertation in 1926. Physical anthropology was then a very small field, and Shapiro was Earnest Hooton's first graduate student. His professional prospects were uncertain. However, when Louis R. Sullivan, assistant curator of anthropology at the American Museum of Natural History, died unexpectedly, Hooton pushed the museum to hire Shapiro as his replacement. Sullivan's work had been on the physical anthropology of Hawai'i, making Shapiro a natural choice to continue his research program. At the museum, Shapiro took over Sullivan's many projects, designing a new "Hall of Man" and cataloging the museum's collection of human remains.

Shapiro also distilled the results of his Norfolk fieldwork into a hundred-page monograph, which Honolulu's Bernice P. Bishop Museum published in 1929. It was the first long-form account of Pitcairn Islanders by a professional anthropologist. However, the text did not focus only on physical anthropology; Shapiro also drew extensively from the vast library of existing texts about Pitcairn and Norfolk Islands, retelling the story of settlement, murder, governance, and eventual transfer to Norfolk in dramatic detail.[34] It irked his editor, Herbert Gregory, who complained that it was all ephemera offering nothing of scientific value. Instead, he argued, Shapiro's measurements comprised his sole worthwhile contribution: "intelligible to few other than professional anthropologists, the tables constitute the essential part; they tell their own story."[35] The islanders' bodies, reduced to anthropometric data, spoke for themselves, he suggested; everything else should be left to popular magazine articles.[36]

Shapiro disagreed.[37] For him, the islands' history constituted the bounding parameters of a living experiment, and even small details like the class status of the mutineers might offer key insights. In Shapiro's treatment of the *Bounty* mythos, we can witness his own developing sense of Pitcairn and Norfolk's status as experimental spaces—we can see the ways the first myth became imbricated or even reliant on the second. "The mutineers and their Tahitian wives brought with them to Pitcairn no idealistic theory for the foundation of a new society," he wrote. Thus, "an account of their life and customs is the story of the development and growth of an unconscious social experiment. . . . That the Pitcairn Islanders were isolated from all contact during the very early years of the colony is a matter of great importance. There are few half caste groups who have no sense of social inferiority due to the stigma attached to half-breeds."[38] History mattered; the island was a "social experiment" as much as a biological one for Shapiro. He thought early accounts demonstrated that Pitcairn and Norfolk were largely isolated from communication with outsiders, and thus free from the influence of racial prejudice that, in Shapiro's view, corrupted the utility of mixed-race populations as anthropological subjects elsewhere. Crucially, for Shapiro, this experiment was also an "unconscious" one, an accident of history. Like many scientists who deployed the metaphor of laboratory or experiment, he focused exclusively on the founding generation and elided a history of repeated colonial contact and control, and especially the islands' long and deliberate construction as managed experimental spaces.

But, if Norfolk Island was now an explicit "experiment" for anthropologists, how did Shapiro understand its results? His assessment, like his imagination of the island as experiment-like in the first place, threaded together scientific and colonial interests in race, reproduction, and "morality." In many ways, his depiction was not dissimilar from those of the colonial administrators who preceded him. He knew the British and Australian reports that chastised the Norfolk Islanders for their "laziness" and "indolence," and he chose to echo their pejorative, colonialist language. Shapiro wrote that the islanders, while physically healthy and capable of performing incredible physical feats, failed to engage in "sustained work"—a deficiency he attributed to the environment of an "island where crops are grown without much effort."[39] He did, however, contest suggestions of moral deterioration. "The morality of the Norfolk Islanders has recently been criticized by various visitors," he wrote. While he thought that the island did not conform to its utopian, nineteenth-century reputation, he rejected the notion that it had devolved into an amoral dystopia. "I do not believe that the unmarried are more immoral than the young men and women of many of the large cities of America and England," he declared, adding, "certainly, if immorality is measured by its sordidity, the Norfolk Islanders can not be said to be essentially an immoral people."[40]

However, the core question for Shapiro, as it was for his administrative predecessors, was whether other forms of "degeneration" had appeared—and especially whether these were the result of inherited "racial" characteristics. But when it came to matters of heredity, Shapiro ran into stumbling blocks. Because he could not obtain what he thought to be accurate family histories through interviews, Shapiro was forced to build his Norfolk genealogies "from publicly available data." Here, the corpus of published texts about Norfolk and Pitcairn was vital; in conjunction with the *Island Register,* it allowed Shapiro to reconstruct much of the community's reproductive past. It is striking that the entire population was so widely written about and observed that an anthropologist could reconstruct their family trees from a library visits. However, public accounts hid any cases of infidelity—which Shapiro regarded as a major problem given his interest in the biological inheritance of racial traits.[41]

Shapiro's racialized interest in "heritage" mirrored that of his administrative predecessors, and similarly led to eugenic "moral" pronouncements. Because Shapiro could not trace inheritance between individual parents and children across generations, he instead considered the entire Norfolk Islander population as a whole. The result was an image of the island more reliant on Hootonian racial "types" than it might otherwise have been. He reconstructed Pitcairn Island's and then Norfolk Island's population numbers from 1789 onward, using these, like Darwin, as an index of their "fitness." He arrayed his many measurements of Norfolk Islanders' bodies in tables, but also treated these as sets. Large statistical aggregates were necessary to overcome Norfolk's deficiency as a laboratory: "Unlike laboratory experimentation, a control cannot be maintained, nor is it possible to measure accurately every factor which enters into a problem such as this," explained Shapiro. "Though the massing of statistics does not palliate a deficiency of accurate knowledge of all the factors involved, nevertheless, it is necessary to make the best of what data exist."[42] Comparing his Norfolk measurements to similar Pacific and European series, Shapiro discovered not only that his islanders expressed traits that were similar to both "parent stocks," but that they often surpassed either the English or Tahitians in measurements of "robustness" and "vitality."

Here, his assessment departed substantially from those of the island's colonial governors. On the question of biological degeneration, Shapiro declared: "I found on Norfolk none of the marks of degeneracy which are said to be prevalent among the present Pitcairn Islanders."[43] Instead, five months of tedious anthropometry on the island and years of statistical cross-comparison in Boston and New York suggested the very opposite. The islanders' mixed-race heritage had granted them a "hybrid vigor" that produced a healthier, more vibrant population. "The crossing of two fairly divergent groups leads to a physical vigor and exuberance which equals if not surpasses either parent stock," Shapiro wrote. "This superiority is not an ephemeral quality which disappears after the F1 or F2 generation, but continues even after five generations."[44] Shapiro's work substantiated Darwin's speculation and reaffirmed Hooton's sense that the inheritance of racial traits could be fruitfully measured in hybrid populations. Crucially, his claims also countered those of more hardline racialists, especially in Germany and the United States. *Science* published a short note on his results, indicating that the mental and physical vitality of the islanders dispelled any notions that either race mixture or consanguineous marriage were inherently deleterious.[45] Melville Herskovits, the famous student of Boas, cited Shapiro's Norfolk study at length in order to assail the racialism of Charles Davenport and R. Ruggles Gates.[46]

However, Shapiro's conclusions by no means blunted hardline interpretations in race science or eugenics. Gates himself asked Shapiro for his Norfolk measurements in 1930.[47] Several eugenicists pointed to Shapiro's Norfolk monograph as one of the most useful studies yet of race mixture, one that proved in conditions as close to a laboratory experiment as possible that racial admixture could be understood and disentangled through the study of living hybrid populations.[48] One eugenicist even used Shapiro's study to encourage people with helpful genes to marry among themselves across multiple generations, to create an incestuous but superior elite.[49] Indeed, popular accounts of Shapiro's research highlighted his work's implications for incest above its implications for the race concept, and even his own article for *Scientific American* emphasized the island's romantic history and present condition rather than engaging too directly in the problem of defining race.[50]

That Shapiro's work did not especially trouble existing racial categories is telling, and is in many ways the result of disjunctures between his dreams of laboratory-like legibility and the realities of fieldwork in a heavily scrutinized space. Because Shapiro could not track the mechanics of racial inheritance between individuals, he was forced to treat the population as a whole, looking for "European" or "Tahitian" types in the aggregate. Shapiro thus maintained the typological racial thinking of his Harvard advisor, who, for his part, celebrated Shapiro's results as "not inferior in interest or importance to Eugen Fischer's classic work" on southern Africa.[51] So far as Hooton was concerned, Shapiro and soon-to-be Nazi race hygenicist Fischer were tackling the same anthropological problems using the same methods in similar field sites, all in the service of the same racial science. That both Fischer and Shapiro inherited not only the parameters of their questions but also their "laboratories" from decades or centuries of colonial governance is one reason why.

HARRY SHAPIRO'S ANTHROPOMETRY ON PITCAIRN ISLAND

Shapiro still longed to travel to Pitcairn, which in his mind remained the most important natural experiment in race mixture on earth. The opportunity would be slow in coming. Instead, Shapiro spent years working at other Pacific field sites. In 1928, he traveled to the Tuamotus and the Marquesas Islands. Between 1929 and 1932, he spent almost a year in Hawai'i, which anthropologist L. C. Dunn termed "a great natural experiment in racial hybridization," like Norfolk on a grander scale.[52] Despite Shapiro's many Pacific journeys, Pitcairn Island remained an elusive, unrealized goal. "My desire to visit Pitcairn, though thwarted, did not die," he wrote, but "it certainly received little encouragement. After each of my successive trips to Polynesia, I sighed regretfully that I was still unable to voyage to that isolated rock in the middle of the South Pacific. Pitcairn seemed like an unrealizable dream."[53]

That changed in 1934, when opportunity presented itself in the form of a forty-nine-year-old millionaire from San Francisco named Charles Templeton Crocker. Heir to a massive railroad fortune, Crocker had recently commissioned a 118-foot yacht, which he christened *Zaca*.[54] He dreamed of using it to command scientific expeditions. "I know little of sailing, the sea rather terrifies me, and I am not insensible to any abnormal motion of the deck," he wrote; it was only "an unaccountable love of adventure" that drove him to the rolling waves of the Pacific.[55] In 1933, the president of the American Museum of Natural History, Henry Fairfield Osborne, suggested that if Crocker wanted "scientific" adventure, it could be had by taking an expedition to collect material for the museum's new Pacific Hall. Crocker agreed, declaring: "Well, here I am, my boat is in San Francisco, and my checkbook is in my pocket; all three are at your disposal."[56]

Though the museum was chiefly interested in procuring bird specimens, the voyage offered opportunities to study human beings, too. The museum's director offered Shapiro as a possible expedition member, calling him a "splendid fellow."[57] After a short meeting, Crocker agreed to bring him aboard.[58] Also on the team were entomologist Maurice Williams, ornithologist James Chapin, illustrator Lee Jacques, photographer Tashio Asaeda, ship's doctor George Lyman, and a dozen sailors.[59] They would journey south to the Marquesas, the Tuamotus, and Tahiti, then east to Mangareva, Rapa Nui, and Chile before sailing northward via the Galapagos. At Shapiro's insistence, Pitcairn was added to the itinerary. It was an ambitious plan, and one that afforded considerable opportunity to harness the southern Pacific as a field for race science. Shapiro declared: "I shall try to measure everything in sight."[60]

The *Zaca* left San Francisco on September 19, 1934; a large crowd gathered to watch it sail out toward the Golden Gate.[61] As the *Zaca* wove its way south through the waves, Shapiro reread Herman Melville's *Typee,* a novel of castaways in the Marquesas he had loved as a boy.[62] Nineteenth-century fantasy met twentieth-century reality as the expedition spent the next three months visiting one island after another; on each, Shapiro collected measurements and blood samples, though his relations with his subjects were often strained. Some demanded money in exchange for access to their bodies, an arrangement Shapiro rejected.[63] He tried sometimes to win friendship and cooperation through his cello, but most audiences were unimpressed.[64] Shapiro, for his own part, was equally uninspired by those he measured. The Marquesans were "not very good looking," he wrote in his field notes, describing them in racist terms as more "mixed" and more "deteriorated" than during Melville's visit a century earlier.[65] He sometimes had more success measuring the dead; Shapiro robbed a number of graves and arranged to have skulls sent back to New York.[66]

On December 20, the *Zaca* left Mangareva for Pitcairn, plying the empty, rolling ocean for two days before the crew sighted the island at 6:30 p.m. on December 22.[67] Shapiro described it in his notes as "a faint gray shape on the horizon." It looked, he decided, somehow smaller than its depiction in his collection of "ancient woodcuts, inky plates, and blurred photographs."[68] It was hard to compete with the outsized image of Pitcairn that a century of mythmaking had generated. As the moon rose, Shapiro looked up and called it "a lovely night to dream about Pitcairn before the disillusion tomorrow"; he knew that disjunctures between utopian dream and embodied reality were likely.[69]

Shapiro's first encounter with Pitcairn Islanders was not unlike that of thousands of other first encounters before it, and would have been familiar to him from travel accounts. Pitcairn's reception of outsiders was well rehearsed, and from arrival to departure, a practiced hospitality would shape Shapiro's scientific investigations. At 7:30 a.m., two longboats launched from Bounty Bay, and soon the *Zaca* was overflowing with visitors as some forty Pitcairn Islanders unloaded packages of curios. Neither scientists nor crewmembers were interested in purchasing souvenirs, but they welcomed the islanders aboard as curiosities in their own right. Shapiro sized up the islanders as he spoke with them, scrutinizing their physiognomy not only with an anthropometric eye, but in the same crass, objectifying language of the colonial officials who had done the same before him. The women he described as "unattractive"; many had lost their teeth, and he thought them "withered and thin with age." The men he deemed stouter and stronger, like "old salts."[70]

Among them, Shapiro spotted faces that seemed uncannily familiar, and realized that these belonged to Charles and Edwin Young, who had allowed the anthropologist Arthur Keith to measure them in London almost two decades earlier. Shapiro knew their anthropometric photographs well. Upon seeing them in person, he decided that they "were the most miserable of the lot," with missing teeth and high-beaked noses, looking something like "what Barnum and Bailey would call The Bird Man."[71] Shapiro approached Edwin, asking him, "Aren't you that man Sir Arthur Keith examined in London?"

"Why yes," he answered. "How did you know?"

"I recognized you from the photograph he published," Shapiro replied.[72] It was a rare instance in which anthropometric photography, a medium intended to transform living subjects into standardized scientific objects, revivified object back into subject. But Pitcairn, as a much-studied space, accumulated studiers in such abundance that recognitions and reexaminations were commonplace.

Common, too, was the routine business of hospitably shepherding arrivals onto shore. After an hour on board the *Zaca,* the Pitcairners guided Shapiro into one of their launches and rowed through the waves toward Bounty Bay. It was a calm day, but approaching the landing was still a dangerous affair; strangers did not know its submerged hazards. In a maneuver that impressed the anthropologist, the oarsmen brought their craft through the surf on a wave and steered swiftly around dangerous rocks. Shapiro had spent the last three months jumping from boats onto shores and gangways, and so refused the arms the islanders extended him at the landing. He leaped to the shore on his own, misstepped, and fell on his face. The islanders lifted him back up; no one laughed.[73] Navigating Pitcairn's physical and human terrain would be no sure thing—but little would be possible without Shapiro's hospitable hosts.

Once safely on land, Shapiro fell more easily into well-rehearsed patterns of hospitality. He elected to live with Burley Warren, whom he described as soft spoken and kind. "The hospitality of Burley" and his wife, Eleanor, "was so warm-hearted, so generous, that I immediately felt at home, and I knew that I should love them," Shapiro wrote. "In their unpretentious way, with the native food and manners of unpremeditated simplicity, they had the art of welcoming a stranger."[74] From the landing, Burley carried the anthropologist's luggage up the Hill of Difficulty and to Adamstown. The settlement itself was nothing like what Shapiro expected from his reading. Paths meandered everywhere. The houses, as he told it, were not set on any particular plan, and the verdant plant life seemed overgrown around them. Beechey, after his 1825 visit, had described a well-ordered English village, or so Shapiro seemed to recall. This was something else, he decided: "There is perhaps too much of the touch of shanty whites about these islanders which make them too close to our seamy side to be truly romantic," he wrote. "One has constantly to be whipping the imagination with scenes from the *Bounty* or with names like John Adams, Thursday October Christian, to keep from forgetting 'romance.' And yet the kindness is very touching."[75] Within his first hours on the island, Shapiro was already describing his experience as a disjuncture between textual utopian mythos and embodied reality.

Shapiro spent his first morning on Pitcairn as a tourist. After breakfast, he accompanied Burley to his "plantation" higher up the island. Walking up the winding clay paths, Shapiro took note of the produce and scanned the village below. Rising above it, he saw the shadowy form of Christian's Cave embedded in a cliff face. After gathering a few pineapples, Shapiro returned to Adamstown and spent an hour wandering the village. He was interested, like most visitors, in relics of the *Bounty* itself. He saw the bronze gudgeon of Bligh's ship and its old vice and anvil. He strolled through the cemetery, spotting familiar names on the tombstones.[76] Throughout his walk, Shapiro read the island against its archive, "silently checking [his] illusions against what [he] was actually seeing." He was not alone in his recourse to texted history. In his conversations with Pitcairners about their ancestors, Shapiro was "told almost to the very phrases the accounts I had read for myself, and I discovered that these modern islanders learned their yarn not from some rich local tradition handed down inviolate through generations, but from the very books I had myself consulted." If anything, Shapiro soon decided that he knew their history in considerably *more* detail; the Pitcairners drew from only a handful of texts, while the anthropologist had read and reread "every tome that twelve years of search had revealed."[77] Even when struck by the differences between legendarium and reality, Shapiro held to the idea that his reading had somehow made him more expert on Pitcairn than its inhabitants.

Discrepancies between textual depictions and his own experience might have given Shapiro pause, but they certainly did not hinder his work.

After all, it was genealogy and anthropometric measurement that would ultimately allow the island to fulfill its promise as a racial laboratory—and it was this that would, in his mind, separate his observations from the touristic gaze of those who preceded him. He would just have to explain himself. He called on the island magistrate, Parkin Christian, to arrange a meeting at which he could explain his work. That evening, the bell in Adamstown summoned the entire community to the meeting house. Shapiro walked the short path from Burley's home by the light of an electric flashlight and stepped into the large wooden building. Inside, a single kerosene lamp lit a hundred faces. Seeing his subjects, Shapiro understood not only his vulnerability, but also the tangle of relationships that separated his "natural" island laboratory from the jars of fruit flies with which he had learned genetics: "Suddenly and unbidden the thought came to me that the success of my work depended on the cooperation of these people, and immediately a horrible fear seized me that they might not understand, might even be hostile and resent strangers, however scientific in aim, prying into their intimate lives. I cursed the personal equation which always confronted me in my studies of the human animal. For a moment I longed for the godlike power of the entomologist or the zoologist who had no need to placate his subjects or to consult their convenience."[78] It was a recognition not only of vulnerability and dependency on his hosts' hospitality, but also of the myriad ways in which Pitcairn was nothing like a laboratory. Even so, he pressed on. Shapiro told the crowd that he was an anthropologist, that he had come to study problems of race mixture, that he would measure them, and that he would trace their genealogies. In his field notes, he reported that their support was "clinched" only when he told them that they would also receive medical care from the *Zaca*'s surgeon.[79]

Shapiro was, by 1934, no stranger to fieldwork—but then, neither were the Pitcairn Islanders. Like Norfolk, Pitcairn had already hosted a series of visits by naval officers and government bureaucrats meant to produce reports about its social, biological, and especially its "moral" condition—and these, too, often fixated on moral concerns over eugenic "degeneration" in order to achieve political ends. After Admiral H. Bury Palliser visited in 1898, for instance, he declared: "There is no doubt that they are deteriorating through the effects of intermarriage and overpopulation."[80] In 1897, the captain of the HMS *Comus* described them as "indolent" and "immoral," and wondered whether intermarriage would "deteriorate" them further.[81] Following eugenically inflected reports such as these, the Admiralty instructed ships' surgeons to investigate the health of the population during later visits. High Commissioner Cecil Rodwell included reports by two medical doctors in the summary of his 1921 visit, which examined the islanders "from the medical and anthropological point of view."[82] One of them recorded their heights, skull shapes, skin tones, hair types, and other racial features.[83] Only four years before Shapiro's visit, Assistant High Commissioner H. G. Pilling visited and produced a report assessing the Pitcairners in racist terms, reporting that "in appearance, the islanders resemble the usual type of Polynesian half-caste to be seen throughout Polynesia. They appeared to me to be of good physique, the majority with bright intelligent faces, but here and there were to be noticed some with the dull bucolic type of countenance usually associated with out-of-the-way villages in isolated districts."[84] A doctor accompanied him, who produced a "physical report" on the children. The consequences of these reports, including disparaging treatments in the press and calls for further colonial intervention, were obvious and consistent.

Accordingly, when Shapiro unpacked his tools in the meeting house the next day, his subjects were very familiar with intrusive examinations like his. As in encounters with doctors during official visits, anthropometric measurement blurred the line between medical care and anthropological scrutiny; across the room, George Lyman set up a free clinic. Crocker noted that Lyman proved much more popular than Shapiro, "as for some reason the islanders were shy about being measured."[85] The anthropologist's method was certainly invasive; he collected roughly the same array of measurements he had on Norfolk Island and followed the same methods, and now also collected blood samples—blood groups, he and other anthropologists thought, might serve as a racial marker and a means to assess racial inheritance in mixed groups.[86] Further blending the expedition's medical and anthropological presence, Shapiro relied on Lyman to "stick the islanders' fingers" for him and was annoyed that the doctor's time was devoted mostly to medical care.

Shapiro conducted most of his work in the meeting house, but took his instruments further afield when necessary. At the school, Shapiro performed a "pigmentation survey," comparing lighter and darker skin tones of the pupils against a standardized scale. He visited some islanders in their homes and chatted with them while measuring their bodies. Vieder Young was old enough to remember the migration to Norfolk in 1856 and told Shapiro about hiding to escape departure on the *Morayshire*'s boats. Shapiro noted the resemblance between Vieder and his sons, Charles and Edwin, though he decided that Keith's two samples were "degenerate copies of the old man."[87] Despite his house calls, at the end of the week, some bodies eluded him, and Shapiro drew up a list of islanders for Parkin Christian to track down—which the magistrate dutifully did.[88]

The experiment that Pitcairn afforded as a natural laboratory was powerful, Shapiro argued, in part because the genealogy of each resident was a known quantity. Parkin Christian loaned him a copy of the *Island Register,* which contained entries for marriages, births, and deaths. Crocker, bored with the island, spent three days copying it for Shapiro.[89] The captain also dispatched some of the ship's crew to the island cemetery, where they copied down inscriptions on headstones.[90] Shapiro corroborated that work by interviewing residents, especially older ones. Novelist James Hall had boasted in *Tale of a Shipwreck* that Mary Ann McCoy could trace most of Pitcairn's genealogy back to the community's founding moments. During his 1931 stay on the island, Hall had asked her about what he euphemistically termed the "puzzling family relationships" on the island, telling her about Shapiro's work on Norfolk. She supposedly said, "Oh, why couldn't he have come here as well? I might have helped him."[91]

A decade later, Shapiro finally enlisted her as an informant. He wrote about her affectionately, referring to her as "Aunt Mary Ann."[92] Her "blindness was provocative of an overwhelming tenderness," he wrote; "I felt that she was as delicate and precious as an ancient porcelain."[93] Conversations with her and other older residents were valuable to the anthropologist, but also reveal how difficult and intrusive his work could be, and how inextricable it was from other forms of scrutiny over the island's reproductive and "moral" lives. Shapiro's visit had caused McCoy serious anxiety, she said. She wanted to help him establish accurate genealogies, but that necessitated revealing secret affairs and infidelities. "I'm so ashamed," she repeated, over and over, before revealing every "illegitimate" birth she could recall. McCoy swore Shapiro to secrecy: "The people will blame me if it came out," she implored.[94] In the face of that revelation, Shapiro broadened his net, asking other informants about cases of birth outside wedlock. Schoolteacher Roy Clark "reluctantly gave [him] the lowdown on most of them." Shapiro's host, Burley, also confessed several "illegitimacies."

Shapiro began to worry that perhaps Pitcairn was even less laboratory-like than previously imagined. "I doubt the genetic value of the islanders," he wrote in his field notes after speaking with his informants; "I am afraid that too much admixture from stray sources and sub rosa promiscuity on the island has botched the record."[95] Far from offering easily legible genealogies threading back six generations, Pitcairn instead evinced the same infidelities and uncertainties over parentage as any other society. Furthermore, Shapiro—like so many investigators prying into the reproductive lives of the islanders before him—became worried that Pitcairn's hospitable receptions were in fact a practiced script, one that deliberately hid its social and biological realities from view. Like the government administrators who preceded him, in the face of disjunctures between the expectation of a legible microcosm and the embodied reality of a society entangled in a complex history of previous encounters, Shapiro instead constructed a paranoid concern over secrecy and duplicity.

At the same time, Shapiro's conversations with his informants are suggestive of just how far the practice of his scientific fieldwork relied on and was imbricated with the island's preexisting forms of hospitality. The Pitcairners were masters in the art of receiving visitors; they welcomed the *Zaca* crew warmly and developed affecting relationships quickly. However, we should not blindly follow Shapiro in reading their hospitality as a duplicitous facade, even if we should understand it as the supportive infrastructure that made fieldwork possible. After all, Ann McCoy, Burley Warren, Roy Clark, and others freely gave out information about histories of marital infidelity to a complete stranger. Rather, it was Shapiro himself who framed his conversation with informants like McCoy as transgressions, moments in which he impolitely trespassed the etiquette of teatime conversation to glimpse secrets lurking below the surface. It was a not uncommon narrative gambit in anthropological writing; to claim access to secret knowledge made Shapiro's data more valuable.[96] Instead, we should understand his reading of Pitcairn's hospitality as a paranoid, sublimated recognition of the islanders' agency in the face of repeated and intrusive colonial scrutiny. Shapiro knew they had good reason to be suspicious of strangers interested in their reproductive lives; he worried incessantly that he could not access verifiable truths and instead only saw rehearsed techniques of resistance to scrutiny. In consequence, epistemic uncertainties engendered by his field experiences would continue to haunt his interpretations, just as they had on Norfolk, and would come to shape his evolving understandings of race and human difference.

"STICKY SENTIMENT" AND INSULAR DERANGEMENTS

Shapiro left Pitcairn Island in the same way as his predecessors, walking down to the landing before climbing aboard a launch, to be rowed through the pounding surf back to his ship. Much of the community came out to the cliffs above—the "Edge"—to wave goodbye. Some three dozen Pitcairners rowed out with Shapiro to bid their farewells, crowding the decks of the *Zaca*. Crocker gave them a few gifts: rice, flour, and some fishing line.[97] They reciprocated with gifts of their own; Shapiro left with a souvenir box, some baskets, and a piece of wood from the *Bounty*'s hull. As they returned to their boats, the islanders sang hymns, a traditional gesture when guests departed the island—though Crocker wished they would stop. They had no ear for music at all, he complained, calling their songs "a fervent manner in which to express sticky sentiment."[98]

"Sticky sentiment" was, however, an apt description of Pitcairn's hold over many of its visitors and studiers. Shapiro's career and his identity remained forever entangled with the spaces and people he visited; his obsession, and the Pitcairners' hospitable accommodation of it, forever shaped his anthropology—and Shapiro remained indelibly marked by the stickiness of his field encounters on Pitcairn for the rest of his career. "I did not discover Pitcairn and am in no way responsible for its existence," he wrote, even if he and the island had become "so closely connected in the public consciousness that to mention the one was to suggest the other."[99] After Shapiro returned to New York, he wrote a popular book about the Pitcairn Islanders, *The Heritage of the Bounty: The Story of Pitcairn through Six Generations,* in which he laid out his vision of race and racial hybridity, his attempt to reshape the scientific world's notion of race and its dynamics.[100] In doing so, he returned even more dramatically to the metaphor of the natural laboratory in order to bolster his claims.

And yet, the sticky and intractable sentiments generated by Shapiro's field experiences haunted his work. The unspoken but obvious disjunctures between dreams of laboratory-like scrutability and the realities of fieldwork in a much-studied colonial space shaped both Shapiro's data and his interpretations of them, with ambiguous consequences for the way he would intervene in arguments about race as a scientific category. Of course, Shapiro did not always reveal the full stickiness of his field encounters in his anthropological texts. *The Heritage of the Bounty,* perhaps unsurprisingly, reveals little of the anxieties that haunt his field notes. But by examining the uses Shapiro made of his data from Pitcairn, by attending to the epistemically deranging force of his fears of artifice, and by searching for the traces of his subjects' humanity in scientific records that sought to reduce them to objects, we can locate much that would otherwise be obfuscated in the liminal distance between fieldwork and publication. Human sentiments stuck stickily to so much of the supposedly objective work of anthropometry.

Let us begin with *The Heritage of the Bounty* itself. It was an ambitious book, combining anthropometric and genetic analysis with a historical account of the island and a narrative of Shapiro's own visit, all rendered in accessible prose. It was by far Shapiro's most successful work, popular with the public and well received by academics. The book made it to Pitcairn, too; Shapiro's host, Burley, wrote to congratulate him on a successful portrayal of the island.[101] Its success was understandable. Not only had novels and films made the *Bounty* story famous again, but the anthropology of the Pacific, too, had achieved renewed cultural prominence. Margaret Mead's *Coming of Age in Samoa* had recently suggested that Pacific islands, as microcosms of the human, could offer a critique of American social forms and mores.[102] Moreover, Shapiro's work spoke directly to issues of racial hygiene embraced by Nazism and still promulgated by eugenicists and segregationists in the United States. *The Heritage of the Bounty* thus thrived on both a renewed American fascination with the Pacific and on its implicit promise to critique the views of racialists.

Arriving at conclusions about race, however, was not easy. Once back in New York, Shapiro began examining his pile of photographs and sheets of anthropometric measurements. These reams of data represented the spans of noses, hips, skulls, and so on of 124 adult islanders and some 60 children. These should have allowed Shapiro to trace the genetic inheritance of traits that he deemed racial. However, especially in light of his conversations with informants like Mary Ann McCoy, Shapiro decided that the data about parentage were so uncertain that classical Mendelian methods would be difficult or impossible. He would not be able to trace traits at the individual level from parent to child. Rather, just as he had done on Norfolk Island, he would analyze his data in aggregate, averaging features for comparison with other populations, and tracking the variability of features between sets of brothers and sisters. The result was once again a recourse to an older mode of anthropology more dependent on typological racial thinking.

In the nineteenth and early twentieth centuries, physical anthropologists often took single specimens as representatives of racial "types," just as Shapiro's mentor Earnest Hooton did. However, an anthropologist conversant with modern statistical methods and the new science of genetics might regard individuals not necessarily as individually representative of racial types, but as unique members of larger populations, aggregations of individuals with unique genetic inheritances. And yet, Shapiro's lack of trustworthy pedigrees prompted a return to typological assumptions. That was already the case even during his time in the field. In his records of individual islanders' measurements, Shapiro placed a small mark in the lower left-hand corner. These served as a record of his subjected assessment of each subject's overall appearance on a scale "ranging from Tahitian to English. The Most Tahitian is T+++, the next T++, and so on to the Intermediate, and then E (English), E+, E++, and E+++."[103] He eventually published a table of those assessments, noting that some islanders could "pass" for English, but none could "pass" as Tahitian (see table 1).

Table 1. Shapiro's racialized analysis

	Strongly English	*Moderately English*	*Slightly English*	*Intermediate*	*Slightly Tahitian*	*Moderately Tahitian*	*Strongly Tahitian*
Males	12	8	13	11	3	3	0
Females	6	5	8	14	13	5	2

Shapiro transferred most of his measurements onto cards and graph paper for calculation. Keeping the sexes separate, he produced averages of the islanders' skin, eyes, heads, faces, noses, and so on. These, too, he ultimately arrayed on spectra between poles of Englishness and Tahitianness.

As a result, Shapiro's uses of Pitcairn Islanders to intervene in the racial debates of his day were ambiguous—and he certainly did not do away with race as a fundamental category of analysis. Indeed, it remained the basis of his analytical approach. However, he was able to declare that claims about the dangers of racial mixture were definitively false. If anything, he proposed, Darwin was correct in suggesting mixture was beneficial. Across *The Heritage of the Bounty,* Shapiro argued that Pitcairners tended in some respects toward their Tahitian ancestors and in other respects toward their English ones, but were clearly vital and healthy, having benefited from "hybrid vigor."[104] They suffered no mental disabilities or diseases. The only deleterious effect the anthropologist could discern was their poor dental health, though he was willing to attribute that to their latent Englishness.[105] In a critique of his own society, Shapiro stressed that, "unlike other half-caste populations" in the United States, the Pitcairn Islanders "never had to eat the bitter bread of social or economic prejudice."[106] Indeed, he closed his section on the islanders' biology by calling up the memory of the people he met on Pitcairn and his relationships with Parkin Christian and "aunt" Mary Ann McCoy, "who possessed qualities of leadership or traits of personality that raised them above the level of their neighbors."[107] The "sticky sentiment" and "human equation" of Pitcairn not only shaped Shapiro's science, they ultimately constituted some of his evidence.

The Heritage of the Bounty was published in 1936, at a time when questions about race mixture grew more salient by the day. The United States had already banned interracial marriage in thirty states, and the previous year, the Nazi regime enacted the infamous Nuremberg Laws. After Franz Boas wrote Shapiro to complain that the Germans were "swamping" scientific congresses as a "means of propaganda," the two soon began working together to develop a response.[108] Their first proposal was to compose a statement decrying Nazi claims, to be signed by leading physical anthropologists. They asked Earnest Hooton to write it, agreeing that the general public would be more receptive and less biased against it than one written by two Jewish intellectuals.[109]

The choice of Hooton was interesting, given that in some respects he and Boas were at the head of very separate anthropological schools founded, in part, on different conceptions of race—but Shapiro was able to triangulate between them. Hooton agreed and soon penned a short declaration titled "Plain Statements about Race." It did not disavow all racialist claims and was weaker than Boas or Shapiro had hoped, but Hooton did argue that there were no "pure" races in the modern world, as all human populations were in his view "mixed"—and studies like Shapiro's had shown them to be healthy and happy. Nonetheless, other scientists, including the Smithsonian's Aleš Hrdlička, the University of Pennsylvania's Frank Speck, and Washington University's Robert Terry, were reluctant to append their names.[110] Ultimately, Shapiro, Boas, and Hooton decided that the last would publish his statement under his name only, in the journal *Science*.[111] Race remained a central and foundational category for physical anthropologists the world over, and few of them were willing to denounce its utility in addressing social problems.

Indeed, at least one Nazi social scientist responded to Shapiro's work by pushing back against claims that Pitcairn, as a racial laboratory, presented proof that race was fluid and race mixture was unproblematic. In 1940, German sociologist Emil Rogner composed a monograph on the social development of the island, citing Shapiro alongside other historical accounts. In his *Pitcairn Island: A Contribution to the Study of the Development of Human Society*, Rogner used the island's history to assert a theory of race mixture and society that, in line with Nazi ideology, rejected both Marxist and classical liberal interpretations.[112] He asserted that the island's early leadership under Adams, strengthened by its adoption of firm religious principles and its tight community spirit, had made its society a robust one. Rogner admitted that the racially mixed islanders were healthy, but he put that down to a fortunate racial inheritance. Like High Victorian anthropologists, he placed "Polynesians" at the middle rather than the bottom of his racial hierarchy, and he noted that the English, while already a "mixed race" themselves, had bequeathed their descendants a set of healthy European characteristics. That said, Rogner argued, there would always be a racial limit to their development.[113]

After the war, antiracist public intellectuals in American anthropology assailed scientific conceptions of race much more forcefully—a project for which Shapiro's Pitcairn and Norfolk work was useful. Pitcairn Islanders appeared in Ruth Benedict's *Race: Science and Politics,* where she deployed them to prove that "new human types" were still forming, as they had throughout human history, and that mixed races flourished.[114] Ashley Montagu, too, drew on Shapiro's work to cite the Pitcairners as a useful example of not only biological but cultural "hybrid vigor."[115] Boas used the islanders for the same purposes in his publications even before and during the war.[116] He consulted Shapiro's Pitcairn data directly again in 1942 while preparing further essays on the question of race and population.[117] Shapiro's Pitcairn work even accrued some citations in American legal scholarship as authors sought to overturn anti-miscegenation laws.[118]

Shapiro himself used Pitcairn and Norfolk to envisage a deracialized future, dreaming of a world that would "become increasingly homogeneous" over time, evolving into mixed-race unity.[119] After the Second World War, Shapiro had opportunities to promulgate such postracial visions. As interdisciplinary pressures from genetics and biology continued to call into question typological understandings of race, and especially after Nazi atrocities had undermined support for eugenics and race science, Sherwood Washburn, another Hooton student, called for a "new physical anthropology."[120] Rather than fixate on "race" as its paradigmatic object of study, it would focus on the evolution of "populations" and even on interdisciplinary analyses incorporating society, culture, and ways of life. Its name would have to change, too; "physical" anthropology had too much baggage, and "biological" anthropology better encompassed its new aspirations. Across the 1940s and 1950s, as Shapiro became a senior member of his profession, he was among the new physical anthropology's most ambiguous adherents. He acknowledged the shift in nomenclature. At the American Museum of Natural History, he replaced his "Hall of Man" with a "Hall of the Biology of Man," and he built a similar exhibition for the New York World's Fair.[121] He served as president of the American Ethnological Society, the American Eugenics Society, and the American Anthropological Association.[122] He used his leadership positions to advance his moderate vision on racial questions; the Eugenics Research Association changed its name to the Association for Research in Human Genetics during his tenure.[123]

Even so, Shapiro's Pitcairners remained flexible and mutable objects of racial evidence in the era of the new physical anthropology, turning up often in the postwar human sciences. Perhaps the strangest place Shapiro's Pitcairn Islanders appeared in that literature was next to a troop of baboons; Frank B. Livingstone compared the early history of violence and murder on Pitcairn with competition for mates among his primate subjects.[124]

Proponents of the continuing utility of the race concept cited them, too—most prominently the University of Michigan's Stanley Garn and the University of Pennsylvania's Carleton Coon, who argued that large aggregates of humans living in relative isolation and passing on heredity traits constituted races more or less in the traditional sense.[125] In a joint 1955 article, they suggested that race was a permeable but nonetheless consistent category, useful for classifying humans groups around common characteristics. There were large "geographical" races in the traditional sense, and "local races" like the Pitcairners, which constituted the lower bound of the category's definition.[126] Pitcairn Islanders, they argued, possessed a unique set of physical characteristics, variable among themselves, but distinguishable in aggregate from other populations—as Shapiro's work had proved. Pitcairn, as a natural laboratory of race, was an ambiguous one.[127]

Shapiro himself never abandoned the utility of the race concept, even as he dreamed of a raceless world. As a pioneer of forensic anthropology, he suggested that categorizing skeletal remains into one of the three "major subdivisions of mankind" was relatively straightforward, even if these populations were internally variable and graded into each other.[128] He also famously joined other physical anthropologists and geneticists to produce a rejoinder to UNESCO's 1950 statement on race, which he and his colleagues regarded as ignorant of physical anthropology's contributions on the subject.[129] Their response in a separate 1951 statement, also published by UNESCO, moderated some of the first statement's claims.[130] Human evolution was complex and dynamic, they argued; social or physical barriers were constantly inducing the formation of separate racial groups, while racial crossing was "constantly breaking down the differentiations so set up."[131] Shapiro was able to make that point especially explicit in a further UNESCO publication on race mixture in which the Pitcairn Islanders appeared as a case study.[132] For Shapiro, race was mutable, contained by flexible boundaries, and dependent on environment—but it was also a "real" category amenable to analysis and description.

Lingering uncertainty over the boundaries and mechanics of race haunted Shapiro from his fieldwork on Pitcairn until the end of his career. His understanding of human differences remained ambiguous, positioned someplace between typological essentialism and a population-centered vision of a post-racial anthropology. Much was involved in shaping Shapiro's thinking—Hooton's racialism, Boas's reforming influence, the rise of new statistical methods and genetic sciences—but the sticky influence of Pitcairn's human terrain surely played its part. The island's hospitality and history of investigation had drawn Shapiro in, and at the same time his recognition of the island's long scrutiny had given rise to a recurring anxiety that certain kinds of knowledge from the island could not be trusted. The result was a science that was simultaneously antiracialist and built on racial assumptions, disproving arguments against interracial relationships even as Shapiro continued to revert back to assumed racial types like "English" or "Tahitian."

And yet, despite the "sticky" human entanglements Shapiro found on Pitcairn, and despite the ways his uncertainty over much of the data he gathered there haunted his analysis of race, Shapiro continued to rely heavily on the metaphor of the natural laboratory when discussing the island and his research. He celebrated the analogy in *The Heritage of the Bounty*, calling it an "unconscious and spontaneous experiment," and suggesting, well after his interviews proved otherwise, that "the biological experiment that blind circumstances have created on Pitcairn offers a rare opportunity for the investigation of the laws of heredity."[133] Descriptions of his work in the *New York Times* did much the same, insisting that "geneticists and anthropologists have sighed for just such a community," asserting that, "if they have their way, the island will remain a kind of laboratory in which nature may be watched as she performs the miracle of welding alien types into an anthropological unit and thus illuminating the dark subject of our own racial status."[134]

Shapiro's insistent language had a purpose. In his foundational history of the borderland between laboratory and field, Robert Kohler suggested that field scientists found themselves in the "shadow" of more prestigious laboratory sciences.[135] For Shapiro, recourse to the metaphor of the laboratory was perhaps less a declarative statement about method and more a bid for recognition and relevance. When he suggested that, on Pitcairn, the dynamics of race mixture "resolved themselves as surely as though they had been active chemicals in a test tube," he was making claims about the veracity and authority of his results, but not the means by which he had achieved them.[136] At the same time, the language of laboratory and experiment occluded the complex human entanglements and human histories—the stickiness—that had both made and made uncertain Shapiro's work. Pitcairn's founding mythology played a major part in his portrayal of the island as laboratory-like, even as he did not regard as problematic the long histories of surveillance and scrutiny that helped to build a global image of the island as exceptional or microcosmic in the first place.

Indeed, no single figure did more to cement Pitcairn Island's status as a "natural laboratory" than Harry Shapiro. *The Heritage of the Bounty* remained in print for decades, republished in 1962, 1968, 1979, and 1986.[137] Classrooms and textbooks used his work to make the islanders a perennial case study for anthropology students.[138] By the 1960s, Pitcairn Island was stodgy canon, and anthropologists could reminisce about a four-field anthropological education that included "a few fruit flies and Shapiro's study of Pitcairn Island" alongside "a little linguistics and archaeology."[139] Pitcairn Islanders, as a fixation of professional anthropology, had completed a massive arc that spanned from the classroom to the Pacific, through scientific and public racial discourse, and finally back into the classroom again—the place where the myth of the natural laboratory did some of its most efficacious work. Indeed, it was in that capacity, on a PowerPoint slide in an introductory lecture in biological anthropology, that I first encountered Pitcairn Island as a student at the Ohio State University in the first decade of the new millennium.

It was only years later, when I encountered Shapiro's subjects in a much different setting, that I found myself struck by the full force of the sticky human sentiment that shaped anthropometric work. I was in the archives of the American Museum of Natural History in New York, thumbing carefully through prints of Shapiro's photographs from Pitcairn Island. I recognized some faces from *The Heritage of the Bounty;* others I had not seen before. People whose names I know from so many governmental reports looked back at me across time. The anthropometric method demanded that subjects look at the camera dead-on, all the better to standardize measurement and comparison. But it also meant that anthropology's human specimens, filed away as objects in cabinets like the one at the AMNH, continue to defiantly meet their viewers' gazes. Adria L. Imada has written that "photographs offered the seductive promise of iconicity and indexicality for settler-colonial officials and lay observers," a colonial technology meant to fix colonial subjects in place. We often speak of photographs "capturing" human subjects, and the multivalent meanings of that verb speak to the hard boundaries and fixed meanings Shapiro sought. However, photographs also defied fixity and objectification. As Imada terms it, "the archive of skin was also what we might call and *archive of kin,*" a counter-archive that "reveals affective approaches to personhood and interdependence" and that shows human connections.[140]

Photographs from Pitcairn and Norfolk Islands, taken by Shapiro and his expedition mates, are archives of skin as well as of kin. Though created in order to fix people as anthropometric specimens, and in a broader sense, in order to fix the scientific meaning of race itself, they record a humanity that defied those boundaries and categorizations. Photographs from the *Zaca* expedition survive in a number of repositories around the world, including the AMNH, the California Academy of Sciences, and the Cambridge Museum of Archaeology and Anthropology. Not all of these are Shapiro's and not all were, strictly speaking, anthropometric; some are by Templeton Crocker, and others are by expedition photographer and artist Tashio Asaeda. They occupy multiple genres, and that they can be read as scientific objects, art objects, touristic objects, and sentimental objects results from not only the ambivalent nature of photography in the colonial field, but also from the ambivalent circumstances of their production. In places where visitors were always reliant on forms of hospitable welcome, anthropometric and expeditionary photographs were touristic and sentimental, too. Their continuing resonances into this century have made that clear. During the last decade, the Cambridge Museum of Archaeology and Anthropology, which holds a collection of photographs from the *Zaca* expedition, has worked to make them available to Pitcairn Islanders. With the images reposted on social media, the children and grandchildren of people the *Zaca* expedition rendered as scientific objects found in these photographs an archive of kin, images of family and friends from almost a century ago.

Here, where I have redeployed them again in the context of an academic monograph in the history of science in the Pacific, Shapiro and Asaeda's photographs remind us that anthropology and anthropologists visited sites like Pitcairn Island only briefly, but did so as iterative entrants into long histories of encounter and reception. Scientific and expeditionary visitors took snapshots in time, attempting to fix meaning in places structured by and in encounters predicated by much deeper histories. The myth of the natural laboratory relied on the possibility of fixity, but even photography could not capture and separate out a human encounter from the larger context that had produced it. Epistemic anxieties and derangements abounded. Though Shapiro, in his words, "cursed the personal equation" that shaped his science, "sticky sentiment" was everywhere. He sometimes hid it, sometimes leveraged it, and, ultimately, read it as a confounding factor that made many of his scientific ambitions untenable—even as it was also a feature of the hospitality that made his work possible in the first place.

CHAPTER FOUR

Legitimating Pasts

Historians and the Sources of Authority

"But let her speak for herself," insisted the editor of an 1829 article in the British *United Service Magazine.* He was introducing one of the first histories of Pitcairn Island ever told, or at least told to outsiders—a narrative of the island's community's past and present offered by Teehuteatuaonoa, one of its founding women.[1] And yet, because Pacific women like her were so rarely afforded the capacity to "speak for themselves" in British accounts, we do not know as much about Teehuteatuaonoa as we might like. The "atua" in her name, referencing divinity, suggests that she came from a privileged station in Tahitian society.[2] We know, at least, that she left Pitcairn in 1817, ultimately making her way home to Tahiti.[3] We also know that the *Bounty* mutineers called her "Jenny," and that the same name was tattooed on her arm alongside the initials "AS" and the date "1789," signifying her attachment to mutineer Alexander Smith (John Adams).[4] Teehuteatuaonoa has often been referred to as "Jenny" in books and articles since; her life is most often invoked in order to tell stories about the lives of Pitcairn's mutineers, who are so often the sole protagonists of the island's histories. From the moment she first appeared in print, Teehuteatuaonoa's biography became the substance from which and on which other names and other stories were inscribed. Despite an insistence that she "speak for herself," her narrative was subsumed by others.

Those who have sought to tell authoritative histories of Pitcairn have relied on stories like Teehuteatuaonoa's, even if she rarely appeared as an agentive figure in them; she became a primary source, not a primary character. From 1814 on, the anglophone world was eager for accounts of the island's missing years, and especially the fates of its mutineer founders.

Teehuteatuaonoa offered an "authentic" perspective on those questions. As the *Bengal Hurkaru* put it, "all other accounts hitherto received respecting His Majesty's ship Bounty, have only been obtained from John Adams," her onetime partner, and readers had reasons to doubt his veracity. A wanted man and the last surviving mutineer of the *Bounty* living on Pitcairn, he "would have his own private motives for deviating from the truth in many interesting particulars, for perverting many facts, and suppressing others." Accordingly, Teehuteatuaonoa deserved "some degree of credit."[5]

Stories like Teehuteatuaonoa's often appear in Pacific histories as "authentic" sources. However, "authenticity" was a category that, in practice, depended on a related but separate category, "authority"—what the literary scholar John Sekora called "begetting, beginning, continuing, and controlling a written text."[6] Teehuteatuaonoa's story was not without its authoritative arbitrators. Her credibility—or at least the "degree of credit," as newspaper editors put it—was established not only by her biography, but by those who retold her narrative in print, reauthoring and legitimating it for their own purposes. As her story circulated and recirculated from the Pacific to the Indian and then Atlantic worlds, editors worked to establish its authenticity, mostly by highlighting the work of authoritative intermediaries who had the social power to adjudicate it as such.

At least two of Teehuteatuaonoa's accounts were ultimately reauthored this way, one in the *Sydney Gazette* in 1819, and the other in *Bengal Hurkaru* in 1826; the latter was republished for an even wider audience in the British *United Service Magazine* three years later. The 1819 account was introduced to readers via correspondence received "by a gentleman of Sydney." Her 1826 account, readers learned, was elicited as an interview conducted by merchant captain Peter Dillon and missionary Henry Nott. The editor implied that Teehuteatuaonoa gave her account in Tahitian, informing us that, like Nott, "Capt. Dillon also speaks the Otahetian language fluently, and has acquired perhaps a more intimate acquaintance with the manners and customs of the South-Sea Islanders, than any other person living."[7] Teehuteatuaonoa's narrative was thus made "authentic," deriving its legitimacy from acts of expert mediation by men deemed to have the authority to judge its authenticity—even if they in turn relied on her authentic narrative to tell their stories in the first place.[8]

In the two centuries since Teehuteatuaonoa told her histories, Pitcairn Island has become the subject of many more historians. Most of them have been outsiders who found others' accounts from or about Pitcairn and used them to compose their own authoritative tellings. The first historian, in some sense, was Alexander Smith, born John Adams, the last surviving mutineer and former partner whose dubious and exculpatory stories Teehuteatuaonoa revised. The most recent, at least at the time of writing, is me, an assistant professor of history at a liberal arts college in the American Midwest, composing these words at my desk. Between us are dozens more, including classics celebrated by Pitcairn and *Bounty* aficionados: Sir John Barrow in 1831, Thomas Boyles Murray in 1854, Lady Diana Belcher in 1870, Rosalind Amelia Young in 1894, Harry Maude in 1968, and Greg Dening in 1994.[9] The full list of historians of the island easily numbers in the hundreds, and indeed, that Pitcairn Island has had more historians than it has current residents is suggestive. We chroniclers and interpreters of the past have found in its small spaces outsized meanings. Like the practitioners of other disciplines, historians have been drawn to Pitcairn's supposed insular exceptionalism, and have played a part in maintaining the myth of the laboratory. Though scholars of colonial and postcolonial history might bristle at the thought, we too have treated Pitcairn Islanders as the subjects of a living experiment—indeed, as I will show in this chapter, the concept of the island as laboratory was central to the foundation of Pacific history as a professional discipline. Even more broadly, all of us, like Teehuteatuaonoa's translators and editors, have used the stories and lives of its inhabitants to tell stories of our own and to make broader claims. That is, after all, the fundamental method of our discipline.

This chapter is a history of some of Pitcairn's historians and the uses they made of the island's past in their own presents, but it is not only that. It is also a history of historical authority, concerned with the ways historians have used stories like Teehuteatuaonoa's to establish not only the authenticity of their claims but also their professional capacity to make them.[10] Pitcairn's status as a supposed living laboratory, as well as its position as the smallest colony of the world's largest empire, are crucial elements in this history of history. Historical and governmental authority were closely imbricated in the Pacific; both were formed in the same context and helped to shape each other. Many of Pitcairn's historians, in particular, looked to its past to make claims about the (de)colonial present and future, and especially to affirm the legitimacy of either its residents or its strangers to rule it—not unlike Teehuteatuaonoa's interviewers and editors, who saw visions of merchant or missionary empires. In so doing, they wove together twinned forms of authority: the right to tell historical truths and the right to govern.

This chapter examines two historians in particular. One, Rosalind Amelia Young, was the first Pitcairner to write a book-length account of her home island, which she published in 1894. The other, Henry Evans Maude, was a foundational figure in the establishment of Pacific history as a professional field during the 1950s and 1960s. Like Teehuteatuaonoa, Young often appears in histories of Pitcairn as a primary source, an authentic voice used by more "authoritative" historians to legitimate their own claims. Maude, on the other hand, appears in histories as a starting point for Pacific historiography, an intellectual forerunner who laid the theoretical and methodological bedrock on which the field of Pacific history was built. However, an attention to their historical modes reveals that Young anticipated in some ways the kind of historicism Maude would later champion, offering an island-centered history that not only relied on the written accounts of sailors and missionaries but also took seriously the oral and material traditions of her grandparents. Sixty years later, Maude would make a similar sensibility the core of the field. Both Young and Maude also derived some authenticity from their own lives and experiences, interweaving the personal and the intellectual. Neither began their writing careers as professional historians; Young used epistolary and missionary networks to find audiences, while Maude began his scholarly career as a colonial bureaucrat serving Britain's Pacific empire.

And yet, though they share a certain kinship, the divergences between the two reveal much about the ways historical authority and governmental legitimacy were imbricated with conceptions of the island as laboratory. Over time, Maude's authority grew and Young's waned for a number of interconnected reasons. Young found in Pitcairn's pasts a vision of local autonomy and a rejection of the utopian or dystopian images so often projected on it by outsiders. Maude, as a former colonial administrator, found in Pitcairn a model for a new Pacific history that he thought would allow technocrats like himself to shepherd the region through the uncertainty of decolonization. He redeployed not only the voices but the historical approaches of authors like Young to pursue that project, even as he consigned them to the archives. Gender mattered here, of course; Maude was above all interested in "Islands and men," as he titled his first collection of essays, and he sometimes pursued those stories by relegating the women of Pitcairn Island to the status of authentic voices rather than authoritative narrators. Nonetheless, for professional historians like Maude, the island laboratory would become a space not only of elucidation but also of disciplinary derangement. Treating Pitcairn as singularly microcosmic allowed him to redefine the parameters and methods of a new Pacific history. However, Pitcairn Island's promise to clarify problems of Pacific governance never cohered; after decades of professional fascination, the model of a postcolonial Pacific that Maude developed would fail him. Far from providing easy answers, Young's and Maude's works both ultimately revealed history's limited power to author legitimating visions of governance, or to arrive at anything other than unsettling and uncertain truths.

ROSALIND AMELIA YOUNG AND THE WOMEN OF PITCAIRN ISLAND

The ability to offer a credible history was a selling point when Rosalind Young published her book on the island's past in 1894. Or, at least, that is what readers were told in advertisements like this one, which appeared in the American Seventh-Day Adventist newspaper the *Signs of the Times:*

> Pitcairn Island, one of the volcanic gems of the Pacific, has been heard of wherever the English language has been spoken. The story of the working out of the problem of human life on its limited territory reads stranger and more thrillingly in many respects than a romance. But most if not all of the tales told and books printed have either been too fragmentary, or incorrect and misleading. It will be interesting to the friends of that miniature world to know that an authentic history has been written, and that by a native of the island, one to the manor born.[11]

Like many invocations of the island before it, this announcement asserted that Pitcairn was not only universally known but also universally legible, a "miniature world" for the "working out of the problem of human life." And yet, it suggested that Young offered something new in a market awash with misleading *Bounty* books and fragmentary Pitcairn tales: her new history was singularly *authentic.* Historical authenticity is an elusive category, to be sure, similar to but separate from such supposedly transcendent and numinous certitudes as *truth,* as well as from such socially freighted categories as *authority.* Young's authenticity, we are told, was underwritten by her identity and position: indigeneity provided her a perspective that brought the reader, if not closer to truth, then certainly further from the mislocations and falsehoods peddled by other, more distant authors.

But we also need not take this newspaper's claim at face value. Historical authenticity is itself a historical category, made, unmade, and remade across time.[12] The authenticity of Rosalind Amelia Young's writing did not just depend on her identity as a Pitcairn Islander, but was buttressed and sometimes threatened by several overlapping identities: she was a devout Adventist in a formerly Anglican colony, an amateur historian in a quickly professionalizing field, and, crucially, a woman writing in a genre dominated by men.[13] Accordingly, her capacity to serve as an authentic source was amplified by her biography, even as her status as an authority would remain contingent for the same reasons.[14]

It might strike us as remarkable that Pitcairn Island should have its own homegrown historian at all, let alone that she should be celebrated by her contemporaries as an authentic one. By 1894, Pitcairn was already among the most written-about places in the world. Like other romanticized colonial spaces, it emerged as an object of tremendous fascination, a thing seen, described, and redescribed by others, but not represented as capable of describing itself. That is doubly true of the island's women, who, like Teehuteatuaonoa, often emerged in narratives about Pitcairn island as a latent, feminized, Polynesian presence lurking quietly beneath its more ostensibly Europeanized surface. They were the wives and daughters of mutineers, not protagonists. The silence of Pitcairn's women as represented objects is nowhere more apparent than in film. Hollywood movies portrayed Pitcairn's female founders as objects of desire whose sexual allure inflamed men to mutiny. The 1956 adventure film *The Women of Pitcairn Island,* released by 20th Century Fox, is the clearest case: sailors arrive on Pitcairn twenty years after the mutiny to discover an island populated only by widows and children who, having now reached adulthood, became objects of desire themselves.[15] In a world awash with texts produced by men, many featuring objectifying fantasies of Pacific women, Rosalind Young's history of Pitcairn Island stands out. She seems to speak for herself.

In what is perhaps the most canonical text in postcolonial studies, Gayatri Spivak famously grappled with the limited capacity of the gendered subjects of empire to speak authoritatively for themselves, at least within the restrictive confines of the colonial project, asking whether such figures ever could produce knowledges that were recognized by the dominant epistemic regimes of Europe.[16] Scholars have grappled with versions of that question ever since, and many accounts of history writing from the colonial margins are concerned with the question of how or whether authors, especially female authors, could speak.[17] In the Pacific, especially, historians have located the authoritative power of women to craft narratives in spaces beyond the page, especially through oral histories or material things; histories live in stories and tapa cloths, not just in dusty, underread scholarly books and journal articles.[18] In the case of Rosalind Young, in no small part due to her supposedly ambivalent racial status, we have a different case: a historian cognizant of her place in the texted order of things, offering an autochthonous vision of Pitcairn history in print. Her career as a historian prompts us to contend not only with the agentive possibilities and limits afforded to women like her, but with the specific visions of the past she crafted within those parameters. A library of books and articles portrayed Pitcairn Islanders as the subjects of an accidental experiment; Young, one of its speaking subjects, pushed back against that notion. Almost singularly, she argued against the idea that Pitcairn or its people were exceptional, neither utopian nor dystopian, but rather a "civilized" community capable of self-governance. To do so, she not only produced historical texts for a wide array of audiences, but also embraced oral and material traditions as part of her narrative.

Rosalind Young's life comes to us largely through her own autobiographical writing. She was born on Pitcairn Island in 1853, and left the island when the community was relocated to Norfolk in 1856. Her father, Simon Young, led the back-migration of several families to Pitcairn eight years later, when Rosalind was eleven years old. The move ended any opportunity for formal education, and she wrote that she would always "regret that it was never my privilege to have gone through some regular course of study, to better enable me to accomplish what has since been my life work."[19] Nonetheless, she read widely from books left by visitors, including such mainstays as Milton, Scott, and Dickens, and she developed a prose style that easily matched the sophistication of the middle-class Victorian women with whom she frequently corresponded.[20] She put her talent for writing to use, and from the 1870s onward devoted herself to two major projects: offering Pitcairn's children the education she herself never received, and representing the island in print for an audience beyond Pitcairn's shores.

There are records of Young beyond her autobiography, too, as might be expected on an island as frequently observed as Pitcairn—and the surveillance under which she lived tells us something of the stakes involved in her work. In 1904, colonial officials in Suva received a letter from Henry Walter Petch, an outsider who had married a Pitcairner. He urged the colonial government not to grant Rosalind Young any additional power as a teacher:

> I know well that upon first acquaintance one gathers the impression that she is above the ordinary, but somehow she falls short and spoils everything she puts her hand to. She is a Pitcairn Islander and like most of them lacks in continuity of system. . . . For anyone who has not lived here it is impossible to even begin to understand the peculiarities of this peculiarly peculiar and irresponsible people. There is much more that might be said against the poor lady and of a much more serious character, as is the case with other *leading* islanders, but to me the fault seems to be that they have degenerated through isolation and from other causes, and are densely ignorant.[21]

The authorities largely ignored Petch, aware that he sought to position himself as a leader of the island and was willing to discredit his neighbors in the process. The letter does, however, reflect the ways the island was perceived during Young's lifetime. Both within and beyond administrative discourse, paranoid suspicions grew that Pitcairn's utopian image was a thin veneer, and that beneath it lurked insidious secrets. In producing her own versions of the island's story, Young was not merely entering into an abstract discourse about the island's supposed "degeneration" or regeneration. Outsiders' assessments of the island's success or failure as a social experiment could have direct impacts not only on the island's governance, but on Young's individual position as the island's teacher.

There was even more at stake than eugenic concerns. In 1886, Seventh-Day Adventist missionary John Tay visited Pitcairn, and he found many converts.[22] The change in religion had serious implications. For nearly a century, Pitcairn had been construed as a bastion of Anglicanism in the Pacific. A shift to Adventism meant a pivot away from Britain and toward the United States. Young was keenly aware of the ways the shift would be perceived by the island's supporters, and even saw evidence of their concerns in print. In her book, she quoted at length from an unnamed "English periodical" whose authors described the conversion as "a matter of regret, therefore, to many who are interested in the little community," and described Young and her neighbors as "simple islanders, ignorant of sophistry and the subtleties of scriptural deductions," who had fallen victim to "the arguments of their fanatical visitor."[23]

Both before and after her conversion to Adventism, Rosalind Young served as a cultural ambassador for Pitcairn. She developed a vast correspondence network that stretched as far afield as Chile, California, Massachusetts, and Switzerland.[24] These relationships often began when she met visitors to the island. A Mrs. Nelson Page vividly recalled one such encounter during her 1878 transpacific voyage. Page's ship approached Pitcairn at 11:00 p.m., and soon a launch emerged from the darkness, accompanied by Young's voice singing loudly across the empty sea. Once Young was on board, Page and her female companions asked her "every question we could conjure up, obtained Miss Y's autograph, and devoted the rest of the time to music." Page described her as a "plump, rather brown, young lady, well-educated, daughter of the minister, and is the school-mistress."[25] Performances of civility and respectability were a long-standing feature of Pitcairn's encounters with outsiders. Music and polite conversation marked Young as "civilized"—even if Page racialized her as "rather brown." In the years after that encounter, Page and Young exchanged gifts by mail, sending packets of flower seeds and penholders.[26]

These material exchanges created lasting, affective bonds between Pitcairn and those who admired it, stretching across oceans and decades.[27]

Letters from Pitcairn Island were sufficiently noteworthy that their recipients often published them in such venues as New York's *Scribner's Monthly* and San Francisco's *Overland Magazine.* It was through them that Rosalind Young first appeared in print, and they eventually allowed her to transform correspondence networks into opportunities to publish her writing. In 1880, Young received a letter from a Massachusetts minister encouraging her to write a history of Pitcairn. He promised to "use the proceeds in procuring goods for this island."[28] Soon, articles such as "How the Pitcairn Islanders Spend Their Sunday" appeared in American newspapers and magazines, though after her conversion to Adventism, Young mostly appeared in that church's periodicals.[29] Young was clearly proud to have achieved some degree of success in print; visitors to Pitcairn soon began reporting that she boasted about her appearances in American magazines.[30] Her writing helped to establish Pitcairn as a legendary place in the Adventist imagination, a site where the church's missionaries had achieved one of their earliest and most lasting conversions in the Pacific—and it was in service of this project that Young wrote her full-length history, *Mutiny of the Bounty and Story of Pitcairn Island, 1790–1894,* published in 1894 by the Adventist church's Pacific Press. It was not only the first book-length account of Pitcairn written by one of its inhabitants, but also one of the first published histories of the Pacific written by a Pacific Islander; it would find a readership well beyond Adventism.

PACIFIC AUTHORSHIP AND THE BOUNDARIES OF HISTORY

Like many historians before her, Rosalind Young placed Pitcairn Island almost at the boundaries of the historical itself. Her book began by situating Pitcairn's local "history" against the grander and ostensibly more consequential pageant of European "History," properly conceived: "Toward the close of the eighteenth century," she wrote, "at a time when events producing the most important results were occurring among some of the nations of the earth, there was being laid, unconsciously, the foundation of a history which in all its points could equal, if not surpass, any tale of fiction."[31] It was a remarkable introductory gambit, gesturing at the synchronicity of the *Bounty* mutiny with the seismic upheaval of the French Revolution—the Estates General convened only a week after Christian rebelled against Bligh in 1789. It also recognized profound differences of scale. That "events producing the most important results" happened in other nations was self-evident. Events gained their significance on Pitcairn very differently. There, the *Bounty* mutineers set in motion a narrative that limned the boundaries separating history and fiction, local and global.

However, Young's narrative departed from that of other historians in important respects. For most writers, events had wrenched Pitcairn well outside the boundaries of "History," producing a heterotopic space apart from but representative of the larger world. Young, on the other hand, pushed against Pitcairn's exceptionalism. Her home was a remarkable place with an unusual history, to be sure, but she portrayed the island as well-connected with and fundamentally part of a broader world. Pitcairn was no insular laboratory, separated from the main. Moreover, even if an "unconscious" act of history had created an experiment, then it was one whose results were already clear. The islanders were neither "saints" nor "savages" in her telling, but people similar in sophistication, both morally and civilizationally, to Europeans or Americans.

Crucially, Young operated both within and beyond the generic confines of historical writing in order to produce her narrative. Hers was a profoundly intertextual book; it not only referenced other books, articles, and letters, but wrote with and against them. That was and remains, of course, a hallmark of historical writing; by the nineteenth century, history had become increasingly codified as a genre ever more rigidly defined by its intertextuality. Historians produced texts *about* the past by objectively interpreting texts *from* the past in dialogue with other histories.[32] Young, however, found authentic sources well beyond the textual archive. She often privileged oral testimony over the written word and reveled in autobiographical subjectivity. The result was a Pacific historicism informed by but also different from that of her Euro-American contemporaries, one that authenticated two separate but related assertions of legitimacy: her own authority to produce narratives, and her community's authority to govern itself.

That is not to say Young did not reference printed documents; she assembled an impressive archive of texts and put them to good use. Some of these were local, such as the *Pitcairn Island Register*, the community's foundational record of births, deaths, and important events.[33] Some were published histories, such as Thomas Boyles Murray's 1854 history of the island.[34] Some were official inquiries, such as British Admiral de Horsey's critical account of his 1878 visit.[35] And some were the products of correspondence. Her 1882 letters with Robert Folger of Massillon, Ohio, are an interesting case. Robert was the son of Mayhew Folger, the captain who "discovered" the Pitcairn community in 1808. Robert sent Young transcribed copies of his father's logbook, from which she quoted at length. But Young's book also quoted Robert Folger directly, using his words to authenticate her legitimacy to author history. Folger pronounced that "there has been a great deal of error and nonsense published" about Pitcairn's past, and that "there are very few living who can enter into the spirit of Pitcairn's history, and . . . a large number of would-be historians are engaged in uttering most senseless pretensions to correct the history of the island."[36] Young thus positioned herself as singularly able to set the story straight, but the reader need not take her word for it.

The book's paratexts, front matter, and back matter do much the same.[37] The appendix of the fourth edition, for instance, features a series of letters written by Young and published in the *New York World,* testifying to the ongoing interest in Pitcairn and Young's authority to narrate it. The book also features an introduction by Edward Harmon Gates, a Seventh-Day Adventist missionary who sailed across the Pacific in a ship named *Pitcairn,* a testament to the church's mythologization of the island.[38] Like many male-authored forewords prefacing women's writing during the Victorian period, his provided a masculine attestation of veracity, one that carried extra import given Young's supposedly ambiguous racial and cultural status. Readers could be assured by Gates that, "as far as his observation goes, [he] believes that the statements contained in this book are strictly reliable."[39] Nevertheless, Gates also made sure to position Young and her community in racial terms. The islanders were kind and hospitable, he said, but also "half-castes, their dark features and black hair plainly betraying their Tahitian blood, though some of them have quite light complexion and blue eyes."[40] To Gates and his readers, there were limits to Young's credibility as a writer. Her racial status made her an authentic source, but at the same time threatened to limit her authority as a historian.

However, what was most remarkable about Young's book—in the literal sense that reviewers remarked on it again and again—was that it sometimes privileged oral tradition over primary texts. Gates emphasized that Young "had access for many years to one at least who remembered events that occurred before the beginning of the present century. . . . She has thus had the best of advantages for obtaining a correct knowledge of the island's history."[41] Here, Gates referred to Elizabeth "Betsy" Young, the daughter of Vahineatua and mutineer John Mills. Born in 1792, Elizabeth was part of the island's first generation and lived until 1883. To Rosalind, she was a cherished storyteller, and her oral testimony, alongside that of other older family members and relatives, helped to reframe stories that were otherwise already well known to readers. Elizabeth remembered her father "sorrowing over the destruction of the *Bounty,* as it was his hope to one day return in her to England," even if it meant hanging.[42]

Young's own grandmother remembered finding the dead body of William McCoy.[43] Elizabeth Young recalled witnessing at "eight or nine years of age" the murder of Matthew Quintal at the hands of Edward Young and John Adams. Rosalind Young's translations of oral narrative into prose retained the subjective experience of her sources. Of Quintal's murder, she described Elizabeth's recollection: "How terrified were all of the little band of women and children who beheld the blood bespattered walls. The dreadful scene was vividly pictured on her mind and memory through the long course of more than eighty years."[44]

In privileging not only a storytelling tradition but also a feminine, Pacific subjectivity, Rosalind Young offered a radical departure from the ways other histories of Pitcairn authenticated truths, and indeed from what constituted the appropriate subjects of historical accounting in the first place. Young's readers saw Pitcairn's past through the eyes of its first and second generations of women. While that gaze was sometimes fixed on the men who were typically the agentive actors in Pitcairn and *Bounty* histories, Young turned her readers' attentions not just toward what women saw or felt, but what they did. Throughout her book, encounters with strangers and changes of political leadership—the stuff of masculine history—were interpolated with ethnographic accounts of life, leisure, and religion. To be sure, many outsiders' travel accounts also obsessed over ethnographic detail. However, Young's deployments were different. For Victorian travel writers and their readers, accounts of the islanders' lifeways were opportunities to assess the community's supposed "regenerative" or "degenerative" progress up and down the ladder of civilization. The basic stuff of everyday life—housing, clothing, farming, praying—was investigated to assess Pitcairn Islanders' capacities to govern themselves. In Rosalind Young's writing, the same questions applied, but her situated autoethnographic perspective offered an unambiguous answer affirming Pitcairners as capable of "civilization" and self-government. She was no outside observer but an interested participant who could leverage her subjectivity in order to celebrate Pitcairn as a creative and self-reliant community, situated in its own web of contacts with the wider world. Her mixture of the ethnographic and the historical to offer an agentive, island-centric perspective in some respects presaged the "ethnohistory" that would eventually dominate professional writing about Pacific pasts almost a century later.

Young's history lingered on what she called "the even, uneventful round of life in the little community," using ethnographic detail to show that Pitcairn Islanders had maintained Pacific traditions while also scaling what her Victorian readers imagined as the ladder of civilization.[45] She described the island's production of tapa, a bark cloth the island's first generations used for clothing, bedsheets, and gifts for visitors. Tapa making was, for the women of Pitcairn's founding generation, a cultural act that communicated the maker's place in a broader Pacific world and a matrilineal heritage, and perhaps Young's multiple references to it attest to its continuing prominence in the late nineteenth century, a period when tapa production was declining elsewhere in the Pacific.[46] At the same time, Young was quick to demonstrate that "gowns, long waisted and bone ribbed, after the patterns sent on shore by ship captains' wives," had overtaken tapa as clothing.[47] Her book's many illustrations and photographs, for example, one captioned "group of native men," show the Pitcairners as well-dressed Victorians.[48] The same process, Young wrote, played out in other realms. The sports and games of her Tahitian grandmothers persisted until the middle of the century, but these too gave way to "civilization." Young relates that the islanders learned to sing hymns, and gained a great deal from that "high and ennobling art."[49] That Young wrote about her island's ways of life in order to respond to other writers' portrayals of Pitcairn and the Pacific is clear—she was a speaking subject who wrote to claw back a degree of dignity from those who would portray her as a degenerated or exoticized object of fascination. Visits by passing ships brought a steady stream of inquisitive guests whose imagination of the Pacific Islands was informed by romantic tropes. "Some questions that have been put to those who go on board ship, even at this late date, are very amusing," she wrote. "Did [she] know what a lump of sugar is?" or "'Can you read writing?'"[50]

She was of course also aware of the appearance of those same tropes in text. Young described an account written by Sir Charles Wentworth Dilke, a British Liberal MP and eventual under-secretary of state for foreign affairs, who toured the "English-speaking countries" in 1866. Dilke famously elaborated a vision of a racially unified "Greater Britain" founded on a common Saxondom that exceeded the boundaries of even the empire itself.[51] He was deeply unimpressed by the Pitcairners, whom he described as "indolent half-castes" constitutionally incapable of doing the hard work necessary to maintain Norfolk Island, and who had fled to an easier life on Pitcairn.[52] His language echoed that of colonial administrators, who mobilized racial and eugenic discourse in order to diminish the islands' political autonomy.[53] An account of his visit to Pitcairn was published in the *Leisure Hour,* and Young tells us it "eventually reached the Island, and was read with much interest by the people." The Pitcairners found particularly amusing an illustration of two islanders, clad only in loincloths, paddling a canoe loaded with coconut leaves and fruits for Dilke's visiting ship—a racialized and exoticized image drawn purely from high-colonial fantasy.[54] The nature of Young's ridicule here is obvious, but it demonstrates well the stakes attached to performances and representations of culture. To demonstrate European lifeways and modes of work was to be deemed capable of political autonomy by Dilke's readers.

Young's writing attended frequently to Pitcairn Islanders' position in the racial order, and she deployed the oral traditions of the island in order to differentiate the Pitcairners from those both she and her audiences considered racially inferior. Chapter 4 of her book, "An Unpleasant Surprise," retells a story from the early reoccupation of Pitcairn after 1859 in order to make those boundaries clear. A group of girls were surveying the abandoned houses of Adamstown when they spotted two strangers in the distance, "one of whom was carrying a gun, the other being a colored man." They screamed and fled, hiding in the forest, but soon "the black man's face appeared through an opening in the trees."[55] They eventually learned, Young tells us, that they were friendly whalers, but her American audience would have understood her meaning when she noted that with the Pitcairn men gone, "their natural protectors were all absent" during the encounter.[56] The anecdote tacitly deployed the threat of sexual violence and feminine victimhood in order to establish the island's essential whiteness. As other scholars have noted, performances of whiteness in the face of racial otherness had a long history on Pitcairn, a rhetorical strategy that placed the islanders on the right side of the colonial hierarchy.[57]

Of course, for late nineteenth-century observers of Pitcairn, the most pressing racial concern was "degeneration," a concept that neatly fused Victorian anxieties around race, eugenics, civilization, governance, and morality.[58] As Young wrote, "the question has been frequently asked whether the people degenerate physically," especially as a result not just of their racial origins but as a consequence of close marriage. She noted that the islanders' front teeth were perhaps more prone to decay than others, but she put that down to environmental rather than hereditary causes. Young's letters attest to her awareness of scientific questions about evolution and degeneration. She repeatedly brought up the topic of Darwin in her correspondence, though she wrote "in a crisp language of no admiration."[59] Young preferred the biblical account of creation.[60] Rather, concerns about the island's supposed degeneration mostly occupied a more basic social and moral register—and it was these that she took pains to correct. Young rejected both utopian and dystopian renderings of Pitcairn as dangerously misguided. "Many who have visited the island have gone away with the impression that the favoured inhabitants breathe a purer air than other people, and an atmosphere wholly untainted by sin," wrote Young. On the other hand, she knew visitors wondered "how it was possible that a people sprung from such a debased stock" could produce a moral society. Her response was a critique of the notion that the island was exceptional at all. "Human nature is human nature the world over," she insisted, "and fallen at that, so that it is certainly a mistake to think that, because so remote from the rest of the world, no vice or sin of any kind mars the character."[61]

It was a message that circulated in multiple editions across the globe. However, the receptions and legacies of Young's work were complex. The British government, or at least the consul in Tahiti and the high commission in Fiji, saw her as a woman worth tracking, a figure who ranked among the island's leadership and whose texts constituted useful intelligence. For the Adventist church, for Pitcairn itself, and indeed for a broad readership across the United States, she was Pitcairn's historian. Her book was adopted in Adventist reading courses, and her account was read widely within the church.[62] The islanders themselves certainly engaged in a brisk trade of the books as souvenirs to passing ships, even after Young herself left Pitcairn to live in New Zealand with her husband, David Nield.[63] After her death in 1924, Nield authored a new edition of her book, ordering two thousand copies, and soon found himself in a debate with his brother-in-law—the island magistrate, Arthur Young—who accused him of ruining the book and profiting from his late wife's legacy.[64] Even today, Rosalind Young's poetry features on island souvenirs, and her portrait has appeared on its postage stamps.[65]

Professional historians like myself have cited Young a great deal; however, we have tended to treat her as a primary source, as raw evidence rather than an antecedent to our own tellings of Pacific pasts.[66] Academic history, as it professionalized, necessarily enforced boundaries between insiders and outsiders—to forge a discipline meant disciplining. Religiously infused with Adventist concerns, Young's vision of the past was in many respects incommensurable with the academy's. But there were other boundaries and incommensurabilities, too. She offered a subjective, feminine, island-centered history, gleaned from the oral tradition of her grandmothers. In those ways, Young approached a Pacific historicity that professional historians would only haltingly begin to embrace a century later.

HENRY EVANS MAUDE AND HISTORIES OF "ISLANDS AND MEN"

Among the historians who would do so was Henry Evans Maude. Though he began his career as a British colonial administrator, by the 1970s, he was a leader in the emerging world of professional Pacific history. In his presidential address to the History Section of the Australian and New Zealand Association for the Advancement of Science in 1971, he announced the field's arrival as a fully formed discipline, striking a triumphant tone:

> Pacific history is not only a fascinating specialization in its own right, studying a regional laboratory of historical variables in miniature that will enable it to make an increasing contribution to the discipline as a whole, but . . . it also has a very practical and therapeutic role to enact in assisting the rehabilitation of the Pacific peoples at the end of a traumatic era of European political, economic and technological ascendency by renewing their self-respect and providing them with a secure historical base from which to play their part as responsible citizens of independent or self-governing communities in a new world.[67]

In a decolonizing Pacific, it was Pacific historians, most of them trained at the department Maude helped to build at the Australian National University (ANU), that would help guide Pacific peoples into a postcolonial order by giving them a "secure historical base" from which to understand their self-governing roles as "responsible citizens." Those new Pacific historians, many of them Maude's disciples, would publish their work in the *Journal of Pacific History*, which he had cofounded six years earlier. They would base their histories on texts made available by the Pacific Manuscripts Bureau, an international system for copying and sharing sources, conceived by Maude and founded in 1968. Perhaps most importantly, they would legitimize their histories by deploying a very old idea: that Pacific islands formed a "laboratory of historical variables in miniature." Maude's vision thus neatly combined the notion of the island laboratory with a call to mobilize history as a legitimating tool of postcolonial governance.

Though Maude was most known for his work on the Gilbert and Ellice Islands (Kiribati), Pitcairn Island played a major role in his project. If the Pacific was a regional laboratory, then Pitcairn was perhaps a laboratory within a laboratory, a site for working out the problem of how to make insular history relevant and meaningful. Several of Maude's first major academic publications, produced after his arrival at the ANU in 1957 and published as part of an edited collection a decade later, focused on the island's early years.[68] In them, autobiography and history quickly merged; before his turn to academe, Maude had spent two decades in the British colonial service as one of its last anthropologically trained administrators. He was the architect of several forced-migration and resettlement schemes—and, crucially, he spent seven months on Pitcairn rewriting its laws, logging more time in Adamstown than any other administrator. Thus, the question of how, precisely, the Pacific would decolonize was a profoundly personal one for him, and in Pitcairn Island, Maude found a microcosm in which problems of Pacific historicism and insular independence were resolved on a miniscule and discernable scale. Its history, he thought, could point not only to the broader Pacific's future, but could serve as a test case for the elucidation of a novel, Pacific mode of history itself.

To make sense of insular, cross-cultural encounters and their meanings, Maude borrowed a term and a method from the Americas: "ethnohistory." If the problem of the Pacific was principally how to understand societies formed from colonial encounter and how to ready their members for self-rule, then it necessitated a new kind of history that could locate in the past an archive of possibilities for the present. Perhaps mindful of infamous dictums that colonized spaces "have no history," Maude was among the first professional historians to search for Pacific pasts both within and beyond text.[69] More crucially, however, Maude suggested that it was not merely different sources that would produce a generative Pacific history but different approaches. The fusion of history and anthropology, of archivally focused historicism and ethnographic attention to culture and meaning, would allow researchers to understand how societies shifted over time, especially when Europeans and Pacific Islanders met on the beach. In the decolonial era, the history of colonial arrivals would inform colonial departures, and allow the right sort of expertly positioned European to guide Pacific peoples into a self-governing future.[70] Maude's colonial experiences, the birth of Pacific history, and emergence of a postcolonial historicity were thus deeply and inextricably intertwined—even as his method only echoed in some ways the approaches Young had pursued seventy years earlier. In particular, Maude found in Pitcairn and Norfolk Islands historically legible examples of migration and settlement, and he would come to view their histories of managed or forced resettlement as models for the Pacific's colonial and then postcolonial future.

Harry Maude was born in 1906, into a family already long shaped by colonial service.[71] One of his ancestors, Thomas Hallett, was a midshipman on the *Bounty*. His immediate ancestors were officers of much higher rank, mostly in the East India Company and the Indian Civil Service;

his father was president of the Legislative Council of Bihar and Orissa. His early years followed the expected path: he attended St. Paul's School in Darjeeling, the "Eton of the East," and then Highgate before going on to Cambridge.[72] Throughout his itinerant childhood, Maude consumed "a steady diet of *The Swiss Family Robinson* and Ballantyne's *Coral Island,* and later on Stevenson, Melville, and Stoddard, with Stewart's Handbook of the Pacific Isles under [his] pillow."[73] Cambridge shaped Maude's romanticism into a vision of the South Seas as a career. He studied anthropology and took a course on the Pacific with Thomas Callan Hodson, himself a former Indian civil servant. Graduating with honors in 1929, Maude applied for the colonial service, purportedly writing "Gilbert and Ellice Islands Colony Only" across his application form. It was accepted, and he soon found himself on a phosphate ship bound for Banaba, then called Ocean Island.[74]

Maude's colonial service was varied. He recruited for the phosphate mines, served as native lands commissioner, and finally rose to resident commissioner. He published ethnographies, working closely with his wife, Honor.[75] Maude's most famous achievement, however, was the managed migration—often coerced—of colonized peoples to new islands. Indeed, Maude has the distinction of setting in motion the last British annexation of new territory, and in turn founding one of the empire's last new settlements. As early as 1931, Maude became obsessed with the idea of settling the then uninhabited Phoenix Islands. "The romance of these little lone islands, lying out under the equatorial sun far to the east, had quite taken possession of me," Maude wrote, and he devoted his time to studying the possibility of a resettlement program, for which he produced detailed reports.[76] With concern growing that the United States would move to claim the islands, the British government assented to Maude's scheme.[77]

Maude conceived of the project as an example of benevolent, liberal imperialism, informed by his own deep study of Pacific history and anthropology. Through the auspices of his official role, he could help the I-Kiribati to realize their historical potential by sailing to new lands, as their own ancestors had done. "The government is under a definite moral obligation to facilitate the emigration," he told his superiors in a 1937 report, insisting that history and anthropology were on his side. He cited contemporary social scientists to suggest that in "virtually all the small islands of the Pacific it has been the practice from time immemorial, whenever the population threatened to overstep the resources of the island, for the surplus inhabitants to set sail in search of new lands." In his vision of empire, then, it was the role of the administrator to assist the colonial subject "in carrying out a custom sanctioned by his forebears."[78]

The Colonial Office, impressed by Maude's detailed memoranda, agreed, and he soon found himself traveling from one uninhabited island in the Phoenix Group to another, alighting on beaches inhabited only by crabs and seabirds to plant the Union Flag and formally claim the land.[79] It was the last time such rituals would play out in the empire. By 1940, several of the islands were home to small communities. Maude was pleased with his work; in an article on the project, he boasted that "we have a thousand peasant proprietors leading happy and contented lives in their own lands, administered by their own island governments, buying their wants and selling their produce in their own co-operative societies."[80] The settlers' economic status was crucial for Maude, and revealed a great deal about how he understood both his role as colonial administrator and the nature of Pacific history. Pacific communities were at a delicate point of transition, he thought. It was the role of figures like Maude to shepherd them to the next stage in their development, which under the logic of liberal empire meant participation in the broader capitalist economy. His vision of the past, here rendered as a history of oceanic voyaging to settle new islands, melded with his vision of a prosperous and semiautonomous future.

The Phoenix Islands Resettlement Scheme helped to cement Maude's reputation. He soon sought to build on it. In a letter to the high commissioner for the Western Pacific, he wrote that "it is hoped to undertake similar projects on a number of other uninhabited islands in the Central and Eastern Pacific, regarding which I have submitted, or am in the process of submitting, detailed reports."[81] Maude was particularly interested in resettling the people of Banaba, the capital of the Gilbert and Ellice Islands colony, and the site of one of the Pacific's most severe instances of phosphate colonialism.[82] The work of extraction was total and left the landscape scarred. The Banabans received some minor monetary compensation, but Maude worried that "it is difficult to see how they can progress as a race," as "life on the dole on an isolated and barren island, in the shadow of a great commercial machine, can have only one ultimate result."[83] Accordingly, in 1942 Maude set in motion plans to purchase the island of Rabi, near Fiji, from Levers Pacific Plantations, hoping to transform it into a new home for the Banabans.[84]

In 1945, after Allied forces drove the Japanese from the Gilbert and Ellice Islands, the British put Maude's plans into effect. The Banabans had been interned by the Japanese on other islands, and the British chose to move them to battle-scarred Tarawa instead of allowing them to return home. Banaba had been effectively destroyed, the British claimed, and reoccupation was not possible. On Tarawa, the Banabans found themselves in a temporary camp, awaiting a move across a thousand miles of ocean.[85] Finally deposited on Rabi, they were given another makeshift camp, where they began building their new lives. Surveying the results of the move in 1946, Maude noted "considerable sickness" and "27 deaths so far."[86] Six months later, he decided the community had failed to progress into economic productivity quickly enough and was "sinking into indolence and apathy."[87] He grew concerned that their sentiments had "turned rather anti-European."[88] Nonetheless, he encouraged the government to stay the course and encouraged the Banabans to remain on Rabi, insisting that "by learning to solve their problems in their own way they will surely develop what they need most at the present time: self-reliance and a sense of responsibility."[89] The condescending, racially and morally inflected language of "indolence" was, as we have seen in chapter 2, part of the common vernacular of colonial governance.

Maude made a name for himself not only as an architect of Britain's newest Pacific colonies, but through his attachment to one of its oldest. Concerned by continuing reports of supposed "moral deterioration," the High Commission for the Western Pacific sought an administrator who could visit Pitcairn Island and produce an updated code of laws. In July of 1940, Maude was ordered to Pitcairn, accompanied by his wife, Honor, and their two-year-old son, Alaric.[90] They were meant to stay for three months—already far longer than any previous governmental visit. Because of the war, three months became seven. The practical substance of Maude's work on Pitcairn mattered a great deal; in it, we can discern the overarching late-colonial vision that not only connected his governance of Pitcairn and his resettlement projects, but that laid the groundwork for his scholarly writing about Pitcairn and the new Pacific history he pioneered in the following decades.

The Maudes arrived on the island as all visitors did, climbing down from a freighter onto a boat before dramatically coming ashore beneath the cliffs at the landing. They already knew that local rivalries would make their work difficult. Members of the Pitcairn diaspora in Wellington had warned them not to trust Nelson Dyett, a "stranger" who worked the wireless transmitter and was locked in a feud with Island magistrate Edgar Christian.[91] Within hours of landing, the Maudes received offers to stay in the homes of both; navigating Pitcairn was "going to be a rather tricky business."[92] Indeed, anxieties over trust and distrust pervaded Maude's initial experience. "I was not unnaturally regarded with a certain amount of suspicion by the inhabitants," he wrote, thinking that they "believed that I had been sent to punish the community and, in all probability, to take away their much prized independence."[93] Faced with a community subject to long-standing scrutiny and colonial intervention, Maude, like others before him, constructed a paranoid vision of insular duplicity. In his official report, Maude insisted that the real Pitcairn was almost inaccessible beneath its hospitable veneer. "The unsatisfactory, at times almost lawless, state of the Pitcairn community was concealed during the first few weeks of my visit by the natural charm and hospitality of the islanders," he wrote, pointing to what he regarded as an "ingrained habit of, to use their own term, 'hypocriting the stranger.'"[94]

Maude's approach to lawmaking was in keeping with his overarching vision of Pacific scholarship and liberal colonial governance. He sought a largely autonomous system, informed by his expert reading of local history, that would foster a community composed of self-reliant individuals. Nothing else would survive Pitcairn's vast distance from any administrative center, he thought. Maude set to work immediately, establishing a council of four island leaders to help advise him in the project of legal reform.[95] They met between August 28 and October 7, 1940, going line by line through the old legal code. Maude looked to the island's records for guidance but found them disappointing. Official government correspondence had been kept in an old sugar bag.[96] Other records were made in pencil and showed visible marks of frequent erasure. Nothing could be trusted, he decided.[97] Instead, like so many before him, Maude privileged outsiders' accounts as more authoritative, especially Harry Shapiro's *The Heritage of the Bounty*. Maude was so grateful for it that he wrote Shapiro a note of thanks later that October, adding: "You would be amused to see us working steadily over the appendix of your book, night after night."[98]

By the end of October 1940, a new constitution was written and endorsed by every adult, native-born resident.[99] Maude ordered portraits of the king and queen and some new Union Flags, meant to symbolize the island's reattachment to the empire. He set up a system for issuing postage stamps, sales of which would keep its government solvent for the rest of the century. In the meantime, the Maudes packed their belongings in boxes in case a ship appeared. As the weeks turned to months, their son, Alaric, began speaking, uttering phrases in the Pitcairn language.[100] It wasn't until April 17 of the following year that, at four in the morning, they heard the signal that a ship would call. Honor described it as a bittersweet moment, writing her mother-in-law to say that she "had got quite fond of them and for many reasons regretted leaving Pitcairn."[101] The island's "sticky sentiment" would keep them attached, however, even at a distance. The Maudes maintained relationships with the Pitcairners for decades, well into Harry's career as a professional historian, when he would begin building his intellectual program by studying and narrating their past.

DECOLONIZATION AND THE FOUNDATIONS OF PACIFIC HISTORY

In 1946, a new Pacific seemed at hand. The war over, Maude returned to Tarawa, the new administrative capital of the Gilbert and Ellice Islands, where he governed the colony as its resident commissioner. But the colonial landscape had shifted. He sensed that Pacific Islanders' "attitude toward the government, and Europeans in general, may have undergone a change," adding that "they are now said to be more openly critical than before, which is ascribed to their having seen the Europeans beaten, if only for a time, by a brown-skinned race such as themselves."[102] To manage their Pacific interests, the major colonial powers set up a South Pacific Commission, and in 1949 Maude became its executive officer for social development. There, he coordinated with academics and technocrats, deploying the same logic of administration that guided his colonial work—namely, the preparation of communities for eventual self-governance. Coerced resettlement, one of the most brutal imperial tools, emerged as a key component of his postimperial vision. In a 1949 paper delivered at the Pacific Science Congress, he insisted that migration was "the only means by which the population pressure in the Central Pacific can be relieved and living standards raised" and that the commission should produce a survey of "the whole South Seas area" in order to find overstretched populations and potential settlement locations.[103] The key to an eventual decolonial future, he thought, was more colonization, guided by an understanding of Pacific history and anthropology.

Maude was not alone in deploying technical expertise in order to manage the Pacific's transition into a postcolonial future. The same year Maude took his position at the South Pacific Commission, James Wightman Davidson accepted the chair of Pacific history at the Australian National University's Research School of Pacific Studies, where he was tasked with building the world's first department of Pacific history. Davidson was a Cambridge-trained historian who understood history as offering a bedrock of technical expertise of use to the present. He had served as an intelligence officer during the war and afterward taught colonial studies at St. John's College, Cambridge. He also advised Western Samoa on constitutional matters and, upon his move to Canberra, worked to build the infrastructure for a new, professional Pacific history that could be of service to the region's governments. Davidson sought like-minded historian practitioners to help build the new discipline, and he thought no one better suited to the task than Maude, whom he persuaded to join the ANU as a permanent research fellow.[104] Together, they would go on to establish what has been called the "Rome of Pacific history."[105]

Maude devoted much of his first decade as a professional historian to Pitcairn Island, treating it as an experimental space in which he could develop his own historicist vision. Even before formally joining the department at ANU, Maude wrote Davidson in 1953 to tell him "I am quite bubbling over with schemes for historical work," and he spent his spare days in Sydney's Mitchell Library locating Pitcairn-related sources.[106] When he formally joined the ANU in 1957, Davidson tasked him with producing a general history of the central Pacific, but Maude decided that the field was too young to support such a project. "It seems clear that as the detailed studies on which a general regional history would normally be based are simply nonexistent," he wrote, "much spade-work has got to be done before such a synthesis can be produced." Instead, Maude began three different Pitcairn projects simultaneously: his constitutional history, one detailing the islands' early years, and one on place-names.[107] By 1958, he told archivist Dorothy Crozier that he would need no end of Pitcairn sources, as he was "writing the most monumental history of that island."[108]

Though the "monumental" history would elude him, Maude did publish considerable Pitcairn and *Bounty* materials. The most substantial of these were the articles "In Search of a Home: From the Mutiny to Pitcairn Island" (1957) and its sequel, "Tahitian Interlude: The Migration of Pitcairn Islanders to the Motherland in 1831" (1959), which tracked the migrations and settlements to and from Pitcairn, as well as a fifty-five-page history of the island in Ross and Moverley's *The Pitcairnese Language* (1964).[109] He also published smaller pieces, including treatments on the island's language, the *Pandora*'s expedition to find the *Bounty* mutineers, and Pitcairn's postage stamps.[110] As the scholarly expert on Pitcairn and *Bounty* history, he reviewed new books and authored encyclopedia articles on the subject.[111] He also continued to provide advice to the British government on Pitcairn; the Western Pacific High Commission asked him to provide "a reliable short history of the island and a reading list."[112] Ever the consummate bureaucrat, he responded with a twenty-thousand-word report.[113]

However, Maude had trouble finding suitable venues for his articles, and his struggles to shepherd his work into print offer a window into the origins of professional Pacific history as a published genre. In 1957, he complained to the Bishop Museum's Margaret Titcomb that "pitifully few journals and institutions are interested in the publication of the results of research in Pacific history, as compared with the enormous number competing for anthropological studies," even through the history was, in his view, "often of considerable practical importance."[114] Accordingly, he and

Davidson launched the *Journal of Pacific History* in 1966; it would quickly become the premier venue for historical scholarship on the region. At the same time, Maude collected his early publications and released them as a book, *Of Islands and Men: Studies in Pacific History,* which opened and closed with his classic Pitcairn papers, and also included a "participant history" of his managed colonization of the Phoenix Islands. Together, these publications constituted a kind of manifesto, announcing the arrival of a new historical specialty.

To justify this new subdiscipline, Maude returned to a very old idea. In his introduction to *Of Islands and Men,* he insisted that Pacific islands were, for the historian, "the counterpart of the natural scientist's laboratory." Scale was important here, he suggested: "The problems presented to the historian are generally no less complex in the Pacific Islands than in other areas. . . . But their occurrence on such an unusually small scale is of great advantage, since it enables research to become far more intensive."[115] It was a logic that explained Maude's fascination with Pitcairn. It offered grand history—colonization, migration, and intercultural contact—on the smallest possible, and thus supposedly most legible, scale. Moreover, as he insisted in a programmatic essay in the *Journal of Pacific History,* Pacific history was laboratory-like because it was especially "recoverable," accessible through written and oral texts extending back across several centuries.[116] With the right academic training and the right sources, the new Pacific historian could offer more exact histories than chroniclers of any other region on earth.

In order to make his case for island history as well suited to exactitude, Maude required a substantial archive—and he had one. Even before his arrival at the ANU, Maude spent much of his career collecting sources, combining a bureaucrat's love of paper and a magistrate's concern with the preservation of testimony. He proudly listed his collection on his CV, boasting that "for over 20 years I have been steadily collecting a reference library of books, pamphlets and periodicals on the Pacific Islands. This by now numbers considerably over a thousand items."[117] While in the colonial service, he kept his library in twenty camphor wood chests, which moved with him around the Pacific.[118] His Pitcairn collection was only a subset of the whole, but it, too, was vast; one of his first projects was the completion of an annotated bibliography of Pitcairn on index cards consisting of some 350 entries.[119] Maude's pursuit of new sources transformed his colonial collecting practices into a new Pacific historical method, one that transcended the then-accepted limitations of texted evidence. He critiqued earlier Pitcairn historians for relying too readily on "the reports or published narratives of a few naval officers," and so conducted an exhaustive search for more material.[120] Maude kept volumes of notes from his days in the colonial service, but also knew where official Pitcairn records were kept and asked for them.[121] He corresponded with librarians and historians in Britain, Hawai'i, Australia, New Zealand, and the US in order to procure previously unconsulted material.[122] He asked anthropologist Harry Shapiro about the journal of Edward Young, attested to in early documents but subsequently lost, a kind of holy grail in *Bounty* studies.[123] While he never found it, he did locate new sources: a manuscript by the mutineer and island patriarch John Adams; the journals of several *Bounty* mutineers; and, perhaps most crucially, the two accounts by Teehuteatuaonoa—or Adams's partner "Jenny"—that opened this chapter.[124]

Using them, Maude distanced himself from his predecessors. To discipline the boundary between amateur and professional historians, Maude had to render his sources as more authentic, and his interpretation as more authoritative. In the first footnote of "In Search of a Home," he dismissed Rosalind Amelia Young and most other historians as "adding nothing to our knowledge of the presettlement period."[125] She had oral tradition only, but Maude had written transcriptions of an oral account, which allowed him to validate, for instance, Teehuteatuaonoa's narrative using Rankean methods of cross-comparison. He read her two accounts against each other and against other sources, such as Adams's autobiographies, and found that "they have proven to be reliable." Teehuteatuaonoa was, he thought, "in her way a remarkable character"—but ultimately a historical subject, someone whose testimony could become the evidentiary substance of a narrative, but who herself was not a historian like him.[126]

The dismissal of a certain kinship with Teehuteatuaonoa or even Rosalind Young as fellow authoritative narrators of the past is not surprising. On the one hand, these *were* primary sources, texts from the past that provided evidence to the historian in the present; I have used them in much the same way. Moreover, the elision of women, especially non-European ones, from the genealogy of the discipline has been well documented as a foundational component of history's professionalization.[127] However, Maude's boundary work here, separating authentic sources from authoritative histories, was complex—not least because of the respect he accorded Indigenous and nontextual sources. He was a pioneer in turning history, as a discipline, toward an expanded archive for narrating the colonial and precolonial past, and collaborated with scholars like Dorothy Shineberg not only to collect material but to write and publish histories.[128] His respect for orality came, he said, directly from his experience as a colonial administrator. Having spent "interminable hours in [I-Kiribati] village meeting houses recounting, savouring, and discussing the traditions of their migrations . . . I came to realize that here was a society possessing an innate historical sense far transcending our own." By the same token, he also insisted that his years working "as a secretariat officer in drafting documentation to suit government purposes" destroyed his former prejudices in the "superior authenticity of the written word."[129] Pacific history's turn toward an expanded set of sources and methodologies was founded, then, in no small part, on the basis of colonial praxis.

The demands of the decolonial era also shaped Maude's embrace of ethnohistory. He saw the need for histories that could "guide" Pacific islanders into independence, and considered Eurocentric accounts of discovery and colonization ill-suited to the moment. Rather, he insisted, it was only "through a proper sense of pride in their history that they will once again find themselves."[130] He demanded that Pacific historians, understood as settlers and descendants of settlers, see themselves as "trustees" working on behalf of Pacific peoples until trained, Indigenous historians could take over.[131] The argument mirrored the patriarchal logic of colonial governance, promising to tutor Indigenous subjects on the path to eventual independence.[132] Moreover, risks of failure, as Maude saw them, were great. Inattention to Indigenous history might mean that Pacific peoples looked instead to "the little Red Book of Chairman Mao," or perhaps worse, that "the work will be done by amateur popularizers of decolonized history" who "over-compensate beyond the bounds of historical prudence in the name of political justice."[133]

Maude's solution was to locate Indigenous Pacific history previously ignored in Eurocentric accounts or deemed inaccessible using European sources, and to analyze them using a marriage of history and anthropology. "The sort of history I am writing is not often easily distinguishable from some branches of anthropology," he wrote in 1960. "In America, I suppose, it would be called ethnohistory."[134] As practiced by the historians Maude mentored, ethnohistory would become a dominant mode for the elucidation of Pacific pasts across the next decades. In Maude's Pitcairn work, islands emerged as either "laboratories" in which cultures met, and only the tools of ethnohistory, applied to the historical archive, could represent both sides of that encounter and make sense of their meanings.[135] The result was often more attention to the island's women than any other historian save Rosalind Young had yet offered. Maude was the first to note, for instance, that Mauatua and Teraura were "of raatira stock—the landed gentry—and thus not inferior in social position to their husbands," who occupied similar stations in British society.[136] Ethnohistorical method thrived alongside the laboratory metaphor; Maude often analyzed Pitcairn as a "unique social and economic experiment" in which novel cultural forms could emerge from the fusion of old ones.[137] He described Pitcairn's houses, for instance, as examples of successful hybridity, thatched in the "Tahitian manner" but framed in a European style. He recounted the island's foodways, its dances, and its language. Maude attended to these cultural forms and innovations in ways few professional historians had before him, and he borrowed from professional anthropology in order to legitimate it. However, Rosalind Young would certainly have recognized his attention to cultural productions and the substance of everyday life.

Ethnohistorical attention to the maintenance or fusion of cultural forms in new environments abetted another, darker feature of Maude's Pitcairn scholarship: its focus on migration. Here, Maude explicitly drew connections between past and present. "It was because I myself organized two successful community colonization schemes, and assisted in the promotion of two others, that I became interested in this prototype of a century before," he wrote in "Tahitian Interlude."[138] As he noted, the British government's forced migration from Pitcairn to Tahiti in 1831 was the first such project in the colonial history of the Pacific, the inauguration of a tradition culminating in his own forced relocations of colonized people over a century later. Even during the 1950s and 1960s, Maude was still consulting with governments on further resettlements. When the British government sought to relocate the people of Nauru, themselves victims of phosphate colonialism, they turned to Maude.[139] He advised that resettlement would succeed only if they secured some degree of "cooperation" and if Naurans were resettled among people similar to themselves. "They could scarcely be expected to intermarry, or even mix into, to any great extent with Melanesians or Australian Aborigenes," he insisted.[140] The same racialized lessons appear in his analysis of Pitcairn history. Maude discovered in the colonial archives that the Pitcairners "never really wanted to leave their island," and only did so under considerable pressure.[141] And, once they were relocated to Tahiti in 1831, Maude blamed the project's failure on the diverging cultures of the Pitcairners and the Tahitians.[142]

Failed resettlement experiments were valuable lessons, Maude thought, and his assessment of them reveals a great deal about the ways he linked colonial pasts and decolonial futures, as well as the racial and colonial assumptions that guided his vision of islands as laboratories. The resettlement experiment in 1831 was certainly a disaster; seventeen people ultimately died. However, Maude thought the brutal experience necessary for the island's historical development, its growth from "adolescence to manhood," in colonialist terms. In escaping the brutality of forced relocation at the hands of the British government, they had been compelled to "assume, for the first time, an adult part in the direction of their own affairs."[143] The patriarchal, developmentalist, and racial logic of British colonialism is obvious here—and it echoes his assessment of the Banabans—as is its radical departure from the humane and women-centered history of Rosalind Young. Even failed migration, in Maude's view, could instrumentalize colonialism as a tool for educating Pacific Islanders in self-government. Maude's vision of Pacific historicity threaded thin and uncertain lines: between ethnohistorical sympathy for those on the other side of the beach than European visitors such as himself and a colonialist vision of subjects requiring tutelage in self-government; between a professional historian's reverence for the written archive and an anthropologist's celebration of oral tradition; between an academic's interest in the past for its own sake and a participant's interest in the Pacific present and future. He helped to forge the professional discipline of Pacific history by stitching together the seams that ran through his own biography.

After 1964, Davidson and others encouraged Maude to abandon his obsession with Pitcairn. Maude requested leave to spend three months working on its constitution and laws, but it was denied. He also found little support for new written work; he told *Bounty* scholar Rolf du Reitz that he still wanted to write at least three more articles on the island, but he complained that "my colleagues are too apt to ask me kindly but trenchantly whether I really consider that the impact of the hundred or two people on Pitcairn Island had such a great effect on the course of Pacific History that I feel justified in spending months writing about their affairs."[144] Maude thought the criticism unfair, not least because it cut to the core of his conception of Pacific history's legitimacy as a discipline. As he told his former colleagues in the colonial service, "the problems of Pitcairn are just as real (and to the social scientist I should have thought fully as important) as those of Fiji, and furthermore they possess the advantage of being probably soluble."[145]

At the same time, Maude received more bad news. The islands of the Phoenix Group, whose settlement he considered the high point of his career, were to be abandoned, the population resettled *again* in the Solomon Islands, on the other side of the Pacific.[146] Maude refused an opportunity to visit the now re-relocated people in the Solomons, worried he would become the "unwilling focus of the pent-up sense of injustice felt by them."[147] Nonetheless, he also considered even this tragic migration "another opportunity which I thought should not be missed . . . to study a group from the very beginning of its resettlement."[148] An anthropologist sent Maude tape recordings he made in the Phoenix Islands in which the people whom he helped to resettle fifteen years earlier recounted their memories of him. He and his colonial ambitions had been consigned to history, archived in the records of another failed migration.[149] Missives like these, Maude wrote, filled him with "gloomy forebodings."[150] A lifetime spent studying the past and present of Pacific migrations, in spaces as small and supposedly laboratory-like in their scrutability as Pitcairn Island, had in the final analysis not made the postcolonial future any more certain.

However, we can nevertheless locate in Maude's and Young's historical work a struggle to intertwine historical and governmental legitimacy, the authority to speak for both the island's colonial past as well as the Pacific's postcolonial future. The differences between Maude's and Young's writings and their receptions were profound, revealing much about the ways historical authority emerged as a category over time, as well as the ways framings of Pacific Islands as laboratories to be studied rather than as sources of authoritative research continue to derange our understandings of authors like them as either "authentic" or "authoritative." That historians like myself do not often remember figures like Young as peers, while we celebrate figures like Maude as founders, is overdetermined. They had differing aims, to be sure—Maude sought an explicit role within the structures of academe, while Young's aims where much bigger. Gender, of course, was part of that story, too. Like so many of Pitcairn's women, Young was often regarded as a primary source, not an author of historical narrative in her own right. Maude, on the other hand, was able to leverage his background as a beachcomber colonial servant into a professional role at the world's most eminent center for Pacific history, where he could offer his accounts "of islands and men." And, perhaps, the notion of the island laboratory or its allied myths as a space of representative exception mattered most of all. In the writing of professional historians like Maude, figures like Young were sources and not narrators—even in historical narratives meant to center Pacific islands and Pacific Islanders' agency.

But perhaps academic Pacific historians like myself should regard figures like Young as forebears in the same ways we do Maude. Despite their obvious differences, the two shared a certain sensibility, and we can locate in both of their bodies of work the origin of intellectual modes that would come to define much of Pacific history in the twentieth century. Young's island-centered history, sympathetic to oral and material traditions and attentive to the culture and ways of life of Pacific people, anticipated the ethnohistory Maude would champion as Pacific history's guiding approach seventy years later. They simply turned it to differing ends. Young's historical vision suggested Pitcairn was no model at all, but rather a community as sacred or flawed as any other. For Maude, the notion of the island as laboratory was key, and even a place as small or marginal as Pitcairn Island could serve as the historical substance from which to build a grand vision of the postcolonial future—even if that de-ranged vision never game to pass.

In that sense, an attention to Pitcairn's treatments as an insular laboratory by my own discipline reveals dark ironies. The field of Pacific history, meant to shepherd the region into a future beyond colonialism, was built not only from colonial antecedents (as so many postcolonial institutions were) but also from one of Pacific colonialism's darkest features: the forced resettlement of people; indeed, for Maude, Young's home and even her own lived experiences were a kind of data emerging from a historical laboratory of forced migration. At the same time, Young pushed back against that colonialist vision and asserted Pitcairn's right to both portray and rule itself, as well as her own right to author its history. Even so, later generations of professional historians, guided by enframing visions of the islanders as laboratory subjects, would relegate her to the role of scrutinized object rather than speaking subject, another primary source whose authenticity professional historians could judge, and which they could use to form and bolster their claims about history or governance—a source of authority rather than an authority herself.

CHAPTER FIVE

Entangled Tapes

Linguists and the Boundaries of Language

You pull an audio reel from its tin, mount it on a player, and attach its end to a receiver. With a hefty mechanical click, you press Rewind. An electric motor whirs, spooling black tape around a plastic spindle until the other end flickers in circles; it has been rewound. You grasp the dangling black thread carefully and pull it through the tone arm, binding it to an empty reel. With another metallic click, you press Play, restarting the machine. The motor moves the reels forward, more slowly this time and more deliberately. Low static vibrates through the speakers until a voice emerges, soon followed by another. You hear two people talking, one in New Zealand English, another in the distinctive sound of Pitcairn's language:

1. M: Good-day to you!
   P: ˌwɒtəˈwɛ·i ˋju | *How are you?*
2. M: Oh, I'm all right. How are you?
   P: ai ˌfi·len ˋsɪkɪ | ai ˋt^{s}u ˋsɔ | ai ˋfi·vʌ | ʻɛs ha· ˋgrrʔp | *I'm feeling sick; I'm very ill; I have a cold; it's influenza.*
3. M: How long have you been feeling unwell?
   P: ai ˈbɪn ˌsɔ· | sɪns ˋjɛsˌtəde | (h)a ˈʔʌdəwʌn ha ˈʔʌdəwʌn | *I've been ill for the past three days.*
4. M: Three days, eh? Are you coming my way?
   P: ö:ˋjes | *Oh yes!*
5. M: Where are you going?
   P: ˈdaʊn tedˋsa·id | *Down to the other side of the island.*

6. M: Down to the other side of the island, eh? What for?
   P: fə ˈeʃ | ən ʔa ˋo·rendʒ | ˌtek ˌa·ʊt ə ˋʃɪp | *For fish, and for the oranges to take out to the next ship.*[1]

The conversation goes on like this for some time, winding around the everyday stuff of life on Pitcairn Island—greetings, illnesses, trade with passing ships.

Interviews like this one were treasured by linguists, who across the second half of the twentieth century took up Pitcairn and Norfolk Islands' shared languages as a subject of vital interest.[2] Like so many investigators before them, linguists sought to understand what seemed to them a complex problem of hybridity. The islands' languages, developing from a founding generation of Oceanic mothers and British fathers, had long struck observers as part English and part Tahitian. Studying them would help scientists to understand contact languages more generally, those pidgins and creoles borne from colonial encounter. Linguists argued that the islands' isolation and well-documented history made them especially useful for understanding how these hybrid languages mixed and evolved over time. As one eminent creolist wrote in 1975: "Pitcairn Island English with its offshoot on Norfolk is of extraordinary interest because it offers as near a laboratory case of creole dialect formation as we are ever likely to have."[3]

This chapter will track the stories of three investigators who took up that idea and made Pitcairn and Norfolk Islands into linguistic laboratories, though there were many others. The first, Albert Moverley, produced the recording above. He came to Pitcairn in 1948 as its first government-appointed teacher; he went on to pursue a PhD in linguistics at the University of Birmingham and, with his advisor, produced the first book on the language of Pitcairn and Norfolk. The second is University of Queensland linguist Elwyn Flint, who conducted the first in-depth linguistic study on Norfolk Island in 1957, collaborating with his informants to produce novel recording practices. The third is Shirley Harrison, a Norfolk Islander who earned a linguistics PhD at Sydney's Macquarie University in 1984 after studying her family's native language during the 1970s. Much of her work was on code-switching: the ability of speakers to reorient from one language to another, often in response to shifts in the situation or context. She became a consummate code-switcher in her own right, pivoting adeptly between the languages of island and mainland, scientist and subject. Taken together, their stories comprise not just a history of language study on Pitcairn and Norfolk, but a situated history of linguistic fieldwork, its practices, and the texts it produced.

Recordings of encounters with Pitcairn and Norfolk Islanders, especially reels of magnetic tape, formed the material basis for that linguistic work. Through them, we can recover not only accounts of linguists' field experiences, but the longer histories of colonial and scientific scrutiny that informed them.[4] Language research did not appear in a vacuum on Pitcairn and Norfolk; long histories of colonial power, strangers' scrutiny, and epistemological anxiety echo across the recordings they produced, subtext to the overt text of informants' voices. This chapter will show how interviewers and subjects together collaborated to articulate new recording practices, treating the islands' longer histories of power, knowledge, and their limits as the resonant spaces in which voices were recorded and interpreted. If one knows what to listen for, those histories are recognizably layered on each tape recording as a background of noise and static.

The resonances and echoes of these islands' pasts as spaces of scrutiny are particularly present in the ways linguists and their subjects collaborated to produce recorded interviews—and in this chapter I argue that they succeeded in producing a new, shared method of fieldwork. In collaborating to produce recordings like the Moverley tape, both linguists and their interlocutors came to the microphone for their own purposes. For linguists, a laboratory-like space was vital for professional recognition. Within the professional study of language, research on "hybrid" languages was long regarded as suspect. Classically, like the philology from which it arose, early professional linguistics was most interested in "pure" languages, mirroring the essentialism of racial anthropology. Pidgins, creoles, and contact languages were discounted as corruptions and jargons; their description was the domain of missionaries and traders, who learned and spoke them for pragmatic reasons.[5] Indeed, so maligned were contact languages that Robert Henry Codrington, who lived on Norfolk Island as the head of the Melanesian Mission in the nineteenth century and who wrote the first substantial study of Melanesian languages, ignored the language of Norfolk Islanders completely and derisively.[6] It was only during the middle of the twentieth century that professional linguists began systematically to study contact languages in sites like Pitcairn and Norfolk. John Reinecke, in some respects the father of modern creole studies, produced his early work on the Hawaiian creole in the 1930s; the discipline grew during the Second World War, knowledge of pidgin languages on Pacific islands suddenly became of considerable strategic interest, and governments sponsored research into them.[7] Building on their wartime experience, American linguists Robert Hall and Douglas Taylor led the development of a new field of contact language studies during the years that followed.[8] By the 1950s, a growing number of linguists entered the new subfield and began conducting studies of sites like Pitcairn and Norfolk Islands, fighting to legitimize the study of these long-stigmatized languages.

Legitimacy was also at stake for Pitcairn and Norfolk Islanders, who found themselves struggling to defend their cultural and political autonomy in the decades following the Second World War. Pitcairn and especially Norfolk came under growing administrative influence, while the concurrent decolonization of neighboring islands put their political futures in question. The islands' demographics were also shifting. Pitcairn's population gradually began to shrink, as young people left for educations and careers in New Zealand. On Norfolk, the "Pitcairn People" found themselves living next to larger and larger numbers of "strangers," mostly Australians—by mid-century, the populations had achieved a rough parity. Among Norfolk's younger generations, use of the local language was dying, in some instances literally beaten out of pupils by government teachers. And so, when Pitcairn and Norfolk Islanders spoke with interviewers, they spoke with men and women who were potential allies in their own struggles for cultural legitimization or political autonomy. But they had given interviews before, and as careful managers of their relationships with outsiders, they knew that outside interest could produce its own undesirable consequences. By the same token, linguists who came to the islands to work often found themselves stumbling over long histories of previous research, by scientists and governmental administrators alike.

Linguistic recordings hum with the static and residual noise of these deep histories and fraught contexts, a feedback loop that grew more powerful with each new interview. Feedback loops figure prominently as a concern in audio recording; when the output of a speaker is fed back into a microphone, resonant noises, reverberations, and room sounds become iteratively reamplified, growing louder and louder. In the most famous demonstration of the effect, avant-garde composer Alvin Lucier famously recorded and rerecorded the same phrase over and over again, until his own voice was occluded by the deafening resonant frequencies of the room itself.[9] A half century of linguistic fieldwork on Pitcairn and Norfolk Islands may have done the same; their spools of tape are inflected by the auditory ghosts and debris of much-longer histories of encounter, amplifying and reamplifying colonial and colonial structures and anxieties. Epistemic derangements followed. Over time, linguists became paranoically fixated on the field encounter itself, aware of the islands' long histories of surveillance and scrutiny, and began wondering whether any recording procedure could access the truth of Pitcairn and Norfolk's languages. In so doing, they also came not only to reassess their own practices, but to scrutinize the boundaries of languages themselves, and to imprint their own concerns over scrutability into their conception of the islands' speech. To locate that history of linguistic anxiety in its longer context of colonial encounter, to listen for this "room noise," let us unspool a few threads of tape, rewind them, and follow their entangled, resonant threads into the past.

ALBERT MOVERLEY AND COLONIAL ANXIETY

The study of Pitcairn's language was born from colonial bureaucracy. In the years after the Second World War, Britain resumed the administration of its Pacific empire, and turned its attention again to Pitcairn. The Western Pacific High Commission had long found the island's Adventist teachers troublesome and sought to appoint an educator of their own.[10] In 1948, they found their man. Albert W. Moverley, who would become the first person to record the islanders' language on tape, would serve as the island's teacher for the next three years. Born in New Plymouth, New Zealand, in 1908, he earned a BA from the University of New Zealand and an MA in history from the University of Otago before becoming a teacher.[11] In 1947 he applied for the position on Pitcairn Island and was chosen over sixty other candidates—it was thought that his experience in rural schools and with Māori students would be applicable to the island's unique circumstances. Moverley, his wife, Jane, and their daughter were told to pack a small number of belongings and travel to Fiji. There, they boarded the *Awahou* already loaded with the prefabricated pieces of a modern school and a schoolteacher's house, as well as three "skilled Indian craftsmen" to assemble them, before sailing on to Pitcairn.[12] They arrived a few weeks later, landing the same way all visitors did on launches crewed by the island's men. The Pitcairners' first interaction with their new government teacher was to load him and his family into their boats, along with the mass of material provided for their house and school. Once ashore, the men winched or carried the building materials from the landing place up the steep Hill of Difficulty to Adamstown, while the Moverleys stood watching with the island's women and children.

Though there to replace the Adventist teachers, Moverley came to the island with a missionary sensibility of his own. In Suva, the high commissioner for the Western Pacific had told Moverley he should act as a "universal guide, philosopher, and friend to the islanders."[13] But the teacher's place in the community was uncertain. In a very literal sense, he operated outside, or even above, the people he came to serve. The new school and teacher's house were built on a hill at the edge of Adamstown, not far from

Christian's Cave, where the mutinous leader had supposedly once looked down upon his new colony in tormented seclusion. As a government report noted, "the average Pitcairn home is roughly built, unpainted, uncared for, and decidedly uncomfortable"—while Moverley's house was large and well-appointed, with multiple rooms, a veranda, hot and cold water, and even a refrigerator.[14] Unlike the established Pitcairn families, however, the teacher had no land of his own to farm—most of his food came from exchanges with islanders. "This type of assistance can be spasmodic however," noted a later teacher, and "they will not accept anything in return which makes it rather embarrassing."[15] The Moverleys' position was delicate in other respects, too. They found friends and allies among a set of Pitcairners who were regarded as having stronger connections with the world of strangers than the world of Adamstown, including postmaster Roy Clark, his wife, Hyacinth, and nurse Evelyn Totenhofer, whom island leader Andrew Young blamed for the death of his adopted daughter.[16]

The Moverleys' relationships with other community members were much more fraught. They especially disliked Andrew Young, whom they called the "island dictator."[17] The Moverleys also developed a fraught relationship with Seventh-Day Adventist missionaries Fred and Myrtle Ward, who had run the island school themselves during the 1930s and had recently returned to Pitcairn. Jane Moverley, who imagined herself as a kind of "social worker amongst the islanders," declared in an angry letter to the Colonial Office that, especially after the Wards' return, Adventism was being used as a "weapon," and that the Pitcairners needed instead a "down to earth guidance which is not so hopelessly beyond the limited heights to which semi-Polynesians can aspire."[18] Albert Moverley wrote critically about the Wards in his own reports, and soon the High Commission refused an extension of the Wards' permission to reside on the island. They were forced to leave Pitcairn in 1951. The high commissioner explained that "experience has shown that the island is too small to contain peacefully more than one stranger in a position of moral authority."[19] Certainly, the islanders understood that they were under surveillance, and that their teacher was reporting on conditions and sentiments. He was powerful enough to send the Wards into exile, at least.

Surveillance by teachers on the island was not only political. Early in his stay, Moverley began assembling a list of local words and phrases. He was hardly alone in taking an interest; across the 1950s, every government schoolteacher who followed him produced some sort of study of the islanders in addition to his government reports. Roy Sanders, who took Moverley's place in 1951, wrote to his superiors in Fiji to announce that he would produce a thesis "on the Behavior Patterns of an Isolated People," emphasizing that "no one here will ever know about it, of course, and the method will by psycho-anthropological."[20] He eventually completed it for an MA in sociology at the University of Auckland in 1953.[21] His successor, George Allen, administered intelligence tests to his pupils to address a long-standing interest in the islanders' mental fitness.[22] The next teacher, Alan Wotherspoon, requisitioned a tape recorder from the Colonial Office "to capture for posterity the unique speech characteristics of these islanders."[23] His replacement, Ernest Schubert, in turn wrote accounts of present-day life on the island for government and academic publications.[24]

Neither was Moverley's interest in Pitcairn's language unique. It had long been an object of fascination and derision, described by visitors either as a curiosity or a sign of the islanders' supposed "degeneration." Colonial administrators treated it mostly as an annoyance. "Conversation I always found difficult," wrote one bureaucrat in 1954.

> Some of the older inhabitants appear to have forgotten the English they once knew (or could not be bothered using it) and used Pitcairnese exclusively in their conversation. The "language" is unintelligible to the outsider. It was mortifying, perhaps in a council meeting, after explaining laboriously some point or other, to find out that the island secretary or chief magistrate would have to turn around and explain it again in Pitcairnese for the benefit of some members. . . . Culturally (in a strict sense of the word) I am inclined to say that with few exceptions the development is not far beyond the moronic stage.[25]

Here, the equation of a creole language with a racialist and eugenic notion that the islanders were unfit to rule themselves was obvious. Other administrators, however, celebrated it as an idiosyncratic feature of island life. In a Reuters article about the island in 1950, Assistant High Commissioner for the Western Pacific Adrian Dobbs described it as "a sort of shorthand English. For instance, what sounds like 'eye kawa' means 'I can't tell you anything about that.'"[26]

Thousands of miles away, at the University of Birmingham, linguistics professor Alan Strode Campbell Ross read that line in the *Times,* or at least that is how he recounted the moment later: "This—to me incomprehensible—remark, and all that might underlie it, seemed to merit linguistic inquiry," he recalled.[27] Indeed, it so aroused Ross's curiosity that he wrote to Dobbs to ask for more information. Dobbs connected him with Moverley, who reported that he was very interested in the language and had already done some work on it. Ross was impressed enough to invite Moverley to begin a doctoral dissertation on what they called

"Pitcairnese." The invitation came at a propitious time. After Moverley spent two years on Pitcairn, the colonial government declined to renew his teaching contract; he had made too many enemies. "Had the Moverleys been more moderate in their approaches," wrote the high commissioner, "the information and advice which they are so anxious to impart would have been welcome."[28] The Moverleys took a steamer home, and the Youngs vowed that there would be trouble if they ever returned.[29]

Before leaving New Zealand for Britain, the former schoolteacher recorded a sample of Pitcairn speech. Wellington was home to a growing Pitcairn diaspora, and it was there that he found an interviewee, a girl of fifteen who had only recently emigrated. They spoke together into a microphone about the island they had both left behind, playacting as though they were still in Adamstown and describing in the present tense a world from which Moverley had been effectively exiled. When they were finished, Moverley took the tape from the recorder and packed it, his field notes, and his family's belongings for the move to Birmingham. There, Moverley met his doctoral supervisor in person for the first time.

Alan Strode Campbell Ross was professor of linguistics at the University of Birmingham.[30] The position was created especially for him in 1951, one of Britain's first chairs specifically in linguistics. He was an expert philologist, working most famously on the Germanic languages. During the early 1950s, he emerged as a widely read expert on markers of class differences in the English, the codes and signifiers that separated the wealthy from the aspirational.[31] He was thus not only a member of a founding generation of modern academic linguistics in Britain, but literally an expert in "proper" English—and so had sufficient credibility to supervise a thesis on what was then still widely regarded as an improper jargon. Moverley began his dissertation under Ross's supervision in 1952. It would be built, in large measure, on the material the erstwhile teacher had brought with him: his notes, a glossary in amateur orthography he had compiled on Pitcairn, and the recording he had made in Wellington.

However, Moverley also brought his political entanglements. He and his wife spent much of the 1952–53 academic year insisting that the Colonial Office intervene more strongly in island affairs—which they described as "moral" and political anarchy.[32] The Moverleys urged the administration to support their friend, island policeman Floyd McCoy, who remained caught in a political struggle with Andrew Young. Jane Moverley also pressed the government to facilitate the divorce and emigration of Alta Warren, who Jane said was being abused.[33] The administration wrote off the Moverleys' critiques as petty annoyance; one administrator recorded that Jane's "attitude for exaggeration is well noted."[34] The Colonial Office made it a policy to purposely avoid her as she was becoming "a source of embarrassment."[35] Unaware of the administration's negative opinion, Ross and Moverley asked the Colonial Office for a grant to fund their project on Pitcairn's language. "I cannot see the least reason why, even if we could, we should contribute funds for Mr. Moverley," wrote one Colonial Office official. "The excuse for this remarkable excursion, i.e. the thesis on the Pitcairnese Language (which I must say sounded to me, when Mr. Young was here, simply like bad English), seems to me most unlikely to be of any practical value to us or to Pitcairn."[36]

Moverley grew increasingly bitter as his relationship with the Colonial Office deteriorated and antagonism with his enemies on Pitcairn increased. In 1953, he complained that a shipment of geological samples he had requested from the island for colleagues at the university had been purposely withheld. In a letter about it to the Colonial Office, he launched into a racist tirade about Pitcairn politics. He was being misunderstood, he said—or rather, administrators had failed to understand the islanders' codes and signals. He urged them to send an official to investigate the island, and that this official bear in mind Pitcairn's languages and practices of obfuscation, as he saw them:

> Let him know that the hallmark of the successful Pitcairner is the ability to "hypocrite the stranger." (Quote is a standard expression.) Let him realize that in a month's stay with ears and eyes wide open all the time he <u>might</u> have a faint glimmering of the <u>truth</u> about Pitcairn. After three years there he would be beginning to get at the basic facts. Let all his statements and commitments be confirmed in writing and made public in his hearing, with nothing left for after his departure, unless he wants his every word perverted. Let his sentences be simple and his words be short. Let him remember that, despite his impressions to the contrary, the Pitcairners are basically and psychologically Polynesian, and that, though they may speak intelligible English, they think in Pitcairnese. Let him not ask negative questions, for the Pitcairn idiom in answering is the opposite of ours, so that "yes" means "no" and vice versa.[37]

That a ranting Moverley should prescribe a field methodology as a corrective to political problems was no accident. The problem of knowing the "truth" about Pitcairn was, for him at least, professional, political, and deeply personal all at once, but his anxieties and experiences cohered in his paranoid vision of the island's language. To understand Pitcairn necessitated understanding informants who, it seemed to him, used language to obfuscate meaning. Moverley, through a racial and colonial lens, thought he had deciphered their codes and significations.

This tirade was, in a sense, one of Moverley's last linguistic works. He died unexpectedly a few months later in 1953, survived by his wife, his daughter, and his unfinished thesis. Ross, still interested in the language, took up the project himself. On the basis of the manuscript and the data his student had gathered, especially the 1951 Wellington tape, Ross set to work writing analyses of his own. His first publication was an article on the island's place-names.[38] He used Moverley's toponymic data to argue that place-names on Pitcairn were "pristine," meaning that the moment of their naming and the conditions leading to it were known historically, as were uses of those names in the time since. As such, isolated but well-documented, Pitcairn constituted a perfect test case for the study of language formation from the moment of its inception, a laboratory for linguistic research. The gap between student's and advisor's assessments of insular legibility is striking—if Ross ever considered the impossibility of discerning the "truth" in the same way Moverley did, he never recorded it. Even so, Ross soon discovered that his "pristine" island would be difficult to access. The data he had at hand were scarce—only a tape, an amateur glossary, and a few notes—and he was not able to undertake fieldwork half a world away. Accordingly, he enlisted the help of a network of linguists and informants around the world. Chief among them was Elwyn Flint, a linguist at the University of Queensland.

ELWYN FLINT AND THE LANGUAGE OF IMPROVISATION

Elwyn Flint was, in many respects, the right person for the job. A linguist who conducted fieldwork across Australia and the Pacific over the course of a four-decade career, he helped to develop the interviewing and recording techniques that would make the study of contact languages possible. Born in 1910, Flint did postgraduate work in English at the University of Queensland before being ordained as an Anglican priest. After his wartime service as a chaplain and intelligence officer, he returned to Queensland as a lecturer, a post he would hold for the rest of his career. Colleagues remembered him as eccentric and "intensely private, manically absorbed in his work."[39] Flint was especially interested in language variation; his principal project was to record languages in Queensland and to document the varieties of English spoken by people across the Pacific, from pidgin languages to aboriginal languages to the dialects of recent immigrants from Europe. In October of 1955, Flint visited Britain and met with Ross, who had just published his first article on "Pitcairnese."[40] As one of the first linguists to systematically tape-record contact languages, Flint was fascinated by Moverley's Pitcairn material. He agreed that considerably more work was needed, and the two became collaborators; however, Flint decided he would collect material not on Pitcairn, but on the more accessible Norfolk Island.

Two years later, in 1957, Elwyn Flint arrived in Kingston with a portable tape recorder. Norfolk Island at mid-century was home to about 300 "Pitcairn People," compared to Pitcairn's 130. They remained colonial subjects; the island was administered as an external territory of Australia, albeit with some forms of local autonomy. Norfolk Islanders circulated occasional petitions urging a reconsideration of their status, but they remained subject to the authority of an administrator appointed by the mainland.[41] The island's changing demographics during the postwar period only added further pressure to the Pitcairn community, members of which began to express serious concerns about the future of their culture and way of life.[42] During the Second World War, Allied forces built an airfield on the island, bringing tourists, foreign residents, and a steadier connection with the wider world in the years that followed. Flint thus entered a field site in flux, a space with a history of intense colonial management and scrutiny, but also one transitioning into an uncertain future, one in which many Norfolk Islanders worried that they would lose their political and cultural autonomy.

Australian colonialism had already done considerable damage to the Norfolk language. As on Pitcairn, administrators on Norfolk had long looked down on it with a mixture of curiosity and derision, regarding it mostly as a corrupt form of English. That did not stop them from recording or describing it. C. R. Pinney, the island's administrator during the 1930s, collected a glossary of local words, writing that it should "give some idea of the limits of expression of the lingo," and demonstrate "the mental handicap the islanders have been suffering from."[43] Another report from 1912 put it even more bluntly, declaring that it was "in no respect a language, it is not even a patois, . . . it is a corruption of bad English spoken by *Bounty* men and imperfectly imitated by the Tahitians."[44] The administration sought its total elimination. In my own conversations with Norfolk Islanders, older informants were quick to tell me about the corporal punishment their teachers meted out when they slipped into their native language.

When Flint and his tape recorder appeared on Norfolk Island, his interlocutors already had some experience dealing with outside researchers, not to mention intrusive, eugenically inflected governmental surveillance. Anthropologist Harry Shapiro had measured and photographed their bodies and produced a glossary of some of their language during his visit in 1923, and American anthropologist Donald Stanley Marshall visited in 1951 to make his own tape recordings. He was warned by the administration that it would be difficult to gain the islanders' confidence, but he recorded in his diary that he "made first contact with the islanders by simple expedient of walking up to house, introducing self, and telling what I wanted—Very hospitably received—they remembered Shapiro with pleasure . . . invited me in for tea." He recorded some of the islanders' speech, which he described as "virtually all English (Cornwall) with little or no Tahitian." His informants were happy to describe their language to him and were anxious to add that "they don't like being called 'lazy' by mainlanders."[45]

Flint also found his fieldwork congenial; like Marshall, he was one of the few researchers of either island who did not resort to aspersions of duplicity or subterfuge. He wrote that he considered the Norfolk Islanders to be among the most pleasant and hospitable subjects he had ever recorded; they were, he said, "highly intelligent, linguistically conscious, and keenly interested in their own language."[46] However, that did not mean Flint found the process of capturing useful samples of their language on tape easy—a long history of scrutiny into the island's affairs and negative appraisals of the language had left a mark. When Flint first began recording, his subjects invariably slipped into standard English, as they had for Marshall. Flint suspected that theirs was a private language; English was used with outsiders. To overcome hesitation would require the improvisation of new field techniques, methodologies produced through collaboration and informed by the island's long history of scrutiny. Ultimately, interviewees began to record themselves speaking without Flint present at all, producing loose scripts to follow in his absence. Their topics ranged from the imaginative to the mundane: the recent centennial celebrations, the island's history of whaling, the appearance of a ghostly apparition, and so on. The tapes they produced are full of stops and starts, of mumbles and hesitant statements as well as brashness and laughter—these were improvised performances.

Let us now take Moverley's Wellington tape off our figurative reel-to-reel player, and instead mount one of Flint's recordings.[47] We press Play, and voices emerge from a speaker. We hear a woman describing chores in the garden (rendered here in English orthography by Flint):

> B: Nex' week I gwen' clean out all dem lantana; I gwen plant der tatie, 'cause ca' do no—no tatie; an er plahn, I gwenner plant der plahn, I gwenner plant some watermelon.

But then, mid-sentence, the tone shifts, and the informant begins to speak more quickly and urgently. The subject is changing. She continues:

> B: We gunna do plenty thing ower ours, because I fraid dah side ower ours ser gone to der pack; an' dah much em horse here, on Norfolk Islan',— You know wha'?—des Islan' gunna come sameas any more side in des worl' . . . Dah–des Islan' ser had it, an' true, we wan' to keep it up, 'cause is ours, an' we wan' to look out good fer des Islan, 'cause, I tell ye what, if nought, dem horse gwenner run us from off des Islan'.
>
> A: Well, I—I agree with you there.
>
> B: Yeah.

A translation of that passage, also by Flint, is as follows:

> B: Next week I am going to clean out all those lantana bushes; I'm going to plant (English) potatoes because one can't do without—without—without potatoes and bananas; I'm going to plant bananas, I'm going to plant some watermelons—we are going to do plenty of things at our place, because I'm afraid our place has gone to the pack. And there are so many of those—outsiders about here on Norfolk Island you know what?—this island is going to get like any other place in this world. That—this island has had it! And true, we want to keep it up, because it's ours, and we want to improve this island—because, I'll tell you what, if not, those outsiders are going to run us off this island!
>
> A: Well, I—I agree with you there.
>
> B: Yes.

Talk of gardening had turned to talk of Norfolk's future, and of the delicate position of the island's "Pitcairn People" in the face of an influx of Australian tourism and migration. Collaboration and improvisation produced a tape recording resonant not only with the sounds of the island's language, but with its history of and frustration with Australian colonialism.

Flint stayed on Norfolk for a month, recording seventeen conversations between thirty-one informants, who ranged in age from young children to nonagenarians. All told, he and his interlocutors produced some 75 minutes of material he deemed usable, or—as the linguist meticulously counted—10,153 words.[48] When Flint returned to Brisbane, he set to work processing these recordings, though even this was begun in collaboration with his informants; while on the island, he produced a loose transcription of his tapes in consultation with his interviewees, like the one excerpted above. Magnetic tape, he knew, produced artifacts and was not to be trusted implicitly—better to corroborate unfamiliar sounds with a text produced in the field.[49] Aided by these notes, Flint transcribed each dialogue into the International Phonetic Alphabet (IPA), transforming audiotape into a standardized text that fixed the sounds of each dialogue in a way legible to every linguist on earth.[50] It was a translation into another medium and, in a sense, a translation into a professional language. Ross also collected every word mentioned in his tapes into a massive form-meaning reference list, "not a linguistic lexicon, but a fully indexed reference list of the linguistic forms and meanings which occur in the dialogues and appendices (elicited material). It includes all variant forms and meanings, and lists function words as well as content words." The list consisted of 155 pages of cross-referenced words, the fixed, categorized substance from which future study could begin. "It is," Flint wrote, "the raw material, not the end product of research."[51] Collaborative and improvised interviews, themselves resonant with the echoes and reverberations of long colonial histories, had now been indexed, codified, and fixed as transferable linguistic data both in text and on tape. These could be shared with linguists across the globe, their entangling threads winding through insular and disciplinary politics alike.

Now, however, came the difficult work of interpreting the tapes' meaning, and here Flint and Ross's interest was centered on two major problems. First, there was the question of how the Pitcairn and Norfolk languages had formed, first from contact and then in supposed seclusion. This was the question that drew linguists to Pitcairn in the first place, and which drove Ross's conception of the island as "pristine." However, any answer required attending to a related set of secondary questions: How *different* were the Pitcairn and Norfolk languages from "English," and what defined the boundary between two languages? A short catalog of the labels applied to the islanders' speech reveals the stakes involved: Was it "speech," a "pidgin," a "patois," a "creole," a "contact language," a "cant," "bad English," an "accent," a "lingo"—or a fully constituted "language" of its own? The answer carried implications not only for the speakers of the language, who would gain a modicum of outside support for a long-threatened language, but also for the linguists who studied them and who wanted to legitimate their subfield.

To answer questions about the boundaries of language, Flint set about determining the mutual intelligibility of Australian English and Norfolk speakers. Following a method used by linguists Donald Laycock and Stephen Wurm in Papua New Guinea, Flint asked colleagues unfamiliar with Norfolk to go through his tape-recorded dialogues "utterance by utterance, and then significant part by significant part of utterance," doing so in sessions of fifty minutes at a time so as "to avoid fatigue."[52] The listener then recorded their own perception of each utterance and meaning on another tape. Flint scored and averaged their results, thus deriving intelligibility quotients for each of his Norfolk dialogues. These ranged from 32.5 percent to 79.7 percent, with an average intelligibility of 57.2 percent. The passage cited above was rated 50.4 percent intelligible. The precise boundary between languages and dialects was understood to be ambiguous; Wurm and Laycock had set it at "around 50%, with the sphere of 40–60% . . . constituting the sub-language range, which is intermediary between language and dialect."[53] Reporting his results at a linguistics conference in Canberra, Flint declared the Norfolk Islanders' speech "a sub-language of English, though some forms of it enter into the dialect range. It is therefore justifiable to call a speaker of Norfolk and English a bilingual." That, in turn, raised new questions: How did these languages coexist and influence each other? What kept them separate? In any case, the ramifications were not lost on Flint. For linguists, the study of communities like Norfolk was "thus no narrow specialist study, but one which directs the attention of the linguist to important practical and theoretical aspects of language."[54] In other words, tape-recorded interviews, themselves redolent with bids for political and cultural legitimacy, in turn helped to legitimate creole studies itself.

Much of Flint's Norfolk work appeared as chapters in Ross's *The Pitcairnese Language,* a comprehensive study of Pitcairn and Norfolk Islands Ross published in 1964.[55] A. W. Moverley, who had died a decade earlier, was listed as its coauthor. *The Pitcairnese Language* was in effect an edited collection of essays by a small army of researchers interested in Pitcairn and Norfolk Islands, including contributions by former colonial administrator Henry Evans Maude, his son Alaric, linguist A. C. Grimson, and onetime Pitcairn schoolteacher Ernest Schubert. It also reproduced transcriptions of tape recordings, including excerpts from Flint's Norfolk dialogues as well as the entirety of Moverley's Wellington tape. Despite the difficulty of producing easy linguistic knowledge on or about either island, especially Pitcairn—which he had never visited—Ross argued that both spaces were vital sites of study for the growing field of creolistics, and models for future work. Echoing early nineteenth-century assertions of Edenic, insular purity, he described them as "pristine." In both islands, "one can witness the actual birth of a language and follow it through to the present day," he declared, adding, "the same kind of thing could perhaps be done for other small isolated communities—only it has not been."[56] Reviewers agreed. One wrote that "if a comparative philologist were to devise a laboratory experiment to test etymological theories, he could hardly devise a better experiment than Pitcairn."[57]

Flint, on the other hand, neither treated Norfolk as a pristine laboratory nor did he describe its people as compliant guinea pigs. In his chapter, he portrayed it as a space subject to considerable governmental intervention and immigration from the mainland.[58] The language, he said, was still active despite government ambitions to eradicate it. He also included long excerpts from his interview transcripts in IPA, accompanied by English translations that were produced by Norfolk Islanders—he would let the voices and meanings of the community itself circulate as knowledge. Remarkably, he also devoted considerable space to a discussion of the unique role his informants played in his field practice. "It may truthfully be said that the people of Norfolk Island are the real authors of anything linguistic written in this chapter," he wrote. "This does not simply mean that the material, representative specimens of which are here presented, is based on tape recordings made and other information supplied by them." Rather, he told his readers, the Norfolk Islanders astonished him with "the breadth of their knowledge and their quick practical grasp of the essentials of what was required . . . so much was this so, that at times there was the strange feeling that the roles of research worker and informants were reversed, and that it was the islanders who were really conducting the project; and this feeling persists."[59] So often, investigators of Pitcairn and Norfolk Islands suggested that their subjects were masters of fieldwork in order to gesture at possible duplicity. For Flint, however, their experience was a virtue—one that, if approached by outside researchers in a mode of earnest collaboration, could prove enormously fruitful for all involved.

Flint maintained an interest in the language and suggested the methods he and his interview partners had codeveloped to other researchers in order to elicit more material. An opportunity came in 1963, when New Zealand's University of Otago sent two teams of social scientists to Pitcairn Island, an archaeological expedition headed by Peter Gathercole and a sociological study led by John Harré.[60] Flint knew that the work of audio recording would be relatively new for both the scientists and their subjects, and he did what he could to facilitate the process. Flint encouraged both teams to let the islanders speak among themselves, without the presence, coaxing, or interruption of outsiders. "The informants tend initially to be shy about using their language to an outsider," he cautioned, "but the Norfolk Islanders found no difficulty in speaking to one another, provided that they were left alone with the tape recorder running, and the research worker withdrawn out of sight and hearing."[61] Flint also sent Harré copies of his 1957 Norfolk dialogues, suggesting that Harré and Gathercole play them for Pitcairn Islanders in order to elicit a response. "Having heard, and presumably understood, what their Norfolk Island kinsmen have been saying and noted their facility of conversation, they might well be ready to give suitable material themselves," Flint advised. Norfolk tapes thus traveled to Pitcairn, where Norfolk Island dialogues, histories, and field methods were, in a literal sense, in dialogue with Pitcairn informants, who in turn produced tapes of their own. Harré ultimately recorded six tapes using Flint's methods, which he duly sent back to Brisbane. His informants were, in many cases, the same people anthropologist Harry Shapiro had measured some three decades earlier. Voices continued to circulate, and resonant histories were amplified as the tapes wound their way back and forth across the Pacific.

SHIRLEY HARRISON AND POSTCOLONIAL FEEDBACK

Flint, Ross, and Moverley were only the first of many linguists to take an interest in the Pitcairn and Norfolk languages. However, those who followed them operated in an increasingly complex political context. During the 1970s, Norfolk Island was once more in the news; its residents were advocating for recognition and greater autonomy against what they saw as unjust incursion by the Australian state. Demographic trends continued, too. A 1976 census showed that of the 859 adults residing on Norfolk Island, only 323 were of Pitcairn descent.[62] Some of them founded the Society of Descendants of the Pitcairn Settlers, whose aims were to "promote knowledge of the Pitcairn race" and to "foster and defend the system of community self-help and self-reliance that developed on Pitcairn Island and exists in Norfolk Island."[63] They sent a petition to the Crown asking that the British government clarify its position on the community's autonomy, and likewise petitioned the United Nations for recognition as a non-self-governing territory.[64] It was in that context that Flint received, in 1979, a letter from a Macquarie University PhD student named Shirley Harrison. She had written a master's thesis, "The Language of Norfolk Island," in 1972 and was now completing a doctoral dissertation on the same subject.[65] Harrison explained that she was researching the language of Norfolk Island and asked Flint for copies of his tape recordings.[66] Flint's dialogues and the field methods that made them possible would thus be taken up by a new generation of linguists—and with them entangled histories of intrusive scrutiny, colonial power, and epistemological anxiety would reverberate again in Norfolk Island's resonant spaces.

Harrison is a noteworthy figure for a number of reasons. Her writing and field research across the 1970s and 1980s constituted at the time the most intense academic investigation of Norfolk Island language to date. Moreover, though a library of books about Pitcairn and Norfolk Islands had been written, their authors were overwhelmingly men. There were crucial exceptions, of course: Mary Russell Mitford was one of its first chroniclers in any form, Diana Belcher became one of its most popular historians, and Pitcairn Islander Rosalind Amelia Young emerged as its most famous native author. Nevertheless, Harrison remains among the few female scientists of any discipline to enter into what was an insular, masculine academic discourse. Perhaps most crucially of all, Harrison's father was a Norfolk Islander, and she was part the islands diaspora. As such, she emerged as a fairly unique figure in the history of research on the islands, an actor who blurred the uncertain boundary between outsider and insider.

Harrison's work was itself built on problems of liminality and hybridity. She became, like Flint, deeply interested in the varieties of Norfolk Islanders' speech, and in the borderlands that separated one language from another. She devoted a decade to studying the islanders' shifts between codes, describing the liminal language that occupied middle grounds between "English" and "Norfolk," or what Flint had come to call "high" and "low" language forms.[67] This last section of the chapter will therefore treat Harrison as an expert on code-switching who herself became an expert code-switcher, a figure who grew adept at navigating and sometimes trespassing the boundaries of several communities at once: island and mainland, insider and outsider, research subject and academic. An attention to her liminal position, and the debates she joined from it, reveals the extent to which the linguistics community, drawn in by both islands' supposedly "pristine" or laboratory-like scrutability, became obsessed with anxiety about islanders' obfuscation and illegibility. At the same time, Harrison's career shows us that notions of her ancestral home islands as either utopically pure or dystopically duplicitous, especially as a scientific field space, were subject to critique by members of the community itself. She was thus part of a wave of researchers from the Pacific who, in the last decades of the twentieth century, joined academic discourses that had been heretofore penned largely by settlers and outsiders.

Shirley Harrison was born on Norfolk Island in 1931, the daughter of Norfolk Islander Moresby Buffett and Australian schoolteacher Mavis Buffett (née Porter). Like most Norfolk Islanders, she knew her genealogy well; she was a fifth-generation descendent of American whaler John Buffett, who settled on Pitcairn in the early nineteenth century. She spent the first ten years of her life on Norfolk before moving to Australia with her family. Though her father spoke the Norfolk language fluently and she heard it a great deal in her childhood, she did not learn it with the same adeptness—a fact that, her field notes suggest, haunted her. Harrison's daughter later recalled that her mother "always understood Norfolk speech quite well but regretted that she did not grow up speaking Norfolk with other children and families."[68] However, her interest in the language remained strong, and even early in her life she sought to study it academically. Harrison first appears in my archive in 1960, when she was twenty-nine years old, stepping into the office of historian and former Pitcairn administrator Henry Evans Maude at the Australian National University.[69] "A dazzling Blonde named Miss Buffett from Norfolk Island, via Armadale, was brought in last week by Wurm and Freeman," Maude wrote in a letter to fellow Pacific historian James W. Davidson. "She wants to do a thesis on the Norfolk Island dialect but I told her of all that was being done by Ross and Flint on the subject."[70] Because of this gendered gatekeeping, it would be almost a decade before Harrison again pursued the project as an MA thesis under linguist Arthur Delbridge at Macquarie University. She would go on to research the language for the next two decades.

Harrison's fieldwork built from Flint's. She was, like him, originally interested in measuring the distinctiveness of Norfolk from standard English, and relied on some of the same techniques he pioneered. And yet, in meaningful respects, she also moved beyond Flint's program methodologically and intellectually. For one thing, Harrison became less interested in the distinctiveness of Norfolk and Australian English, in the hard boundaries separating the island from the wider world, and instead began to study the fluid spaces between the two. In her dissertation, especially, she no longer simply studied "pure" Norfolk language, what she called "Broad Norfolk," or its differences from the English spoken on the island, which she called "Norfolk English"—but rather the continuum of expressions that ranged between them, which she called "Modified Norfolk." For another, unconstrained by the notion of Norfolk as a pristine laboratory isolated from the mainland, she began her work by interviewing the Norfolk Island diaspora in Australia. Deranged by their obsessions with the islands as laboratory-like spaces of exception, other researchers had largely neglected the diasporic community. Moverley had recorded a Pitcairn Islander in Wellington, it was true, but it was a context that the recording itself worked to hide by playacting imagined scenes on Pitcairn.

Harrison's own father became one of her major informants. He was born on Norfolk Island in 1904 and remembered life before the airfield brought in its steady stream of outsiders, when whaling, fishing, and farming still ruled Norfolk's economy.[71] Buffett spoke the language adeptly, though he actively discouraged his own children from learning it.[72] His attitude grew more positive in his later years, when his daughter brought a tape recorder into his home and began recording his speech. Harrison eventually based much of her understanding of Norfolk's grammar, phonology, and lexicon on interviews with him. Building on his networks of kinship and acquaintance, she would go on to interview dozens of other Norfolk Islanders living in Australia. Eliciting natural speech proved difficult, however, and she began bringing her father to her interviews, deploying him as a kind of experimental "constant" whose steady presence would bring order and regularity.[73]

Like Flint, Harrison found that the politics of language and histories of scrutiny were not easily overcome. After one interview with an older Norfolk Islander in Sydney, Harrison lamented that her informant "normally speaks N as Broadly as anyone," but "preferred not to commit herself to N on tape," and would not "acknowledge her competence in N while being taped."[74] Like Flint, Harrison addressed the islanders' reluctance by adapting her approach. She developed a pattern of conversation topics that proved conducive for eliciting "broader" forms of the language. Those "associated with local affairs," especially the "arrival of ships and planes" and "criticisms of island officials, administrators, school master, doctor, etc." were particularly generative.[75] Harrison also administered her interviews in a well-planned order. First, she would ask about specific words or constructions. Only then would she try to force her interlocutors away from the familiar, through the introduction of "variant actors in the situation that might lead the addressees away from [Broad Norfolk]," to see how their speech changed. She found it useful "if a non-islander can come and go while the tape is being made to see effect of his presence, effect of the addressee speaking to him," she wrote.[76] Like Flint advised, Harrison often found it best to simply remove herself from the room when possible. In her absence, the islanders would improvise their own dialogues.[77] Harrison was deeply cognizant of the difficulty of capturing so ephemeral a quality as code-switching. "[It] is likely that insufficient evidence would be available to test this from my tapes," she wrote in her field diary after completing her first twenty-seven interviews. "Note the difficulties of recording islanders' natural behaviors in this regard since it is a delicate point of speech behavior and awareness that he is being observed and recorded is likely to affect the speaker's response."[78]

Harrison brought these field techniques to Norfolk Island itself, in a series of across the 1970s and 1980s. Her notes suggest that she experienced a number of anxieties when working on the island; she worried over her status as an outsider and her own capacity to switch between codes. After one visit to the school, she told an informant that she had "difficulty at eliciting Norfolk from Norfolk Islanders when I speak English to them."[79] She told another informant that her inability to speak fluent Norfolk "prevents me from being able to tape my own conversations with Norfolk Islanders."[80] Harrison had no trouble translating from English to "Broad" Norfolk for other linguists, of course, but speaking fluently on the island was another matter.[81]

Harrison made many recordings on Norfolk Island, often deploying the improvisational techniques Flint pioneered while also developing new ones. She wrote the headmaster of the island school to ask for permission to record his pupils. "Interview is too formal a word, I feel," she told him, explaining her practices. "I would ask them to discuss (in Norfolk) those subjects that would present no difficulty in order that I might listen to and tape the conversation so that I would have it for my later experience."[82] The principal assented, and Harrison made several visits, recording its Norfolk-speaking students in groups of about a half dozen, eventually collecting dozens of interviews.[83] She was especially interested in recording the speech of students who not only came from "pure" Norfolk families, but from mixed households like her own. When not recording formal interviews, she wandered the island listening to informal conversations. Descriptions of surreptitious eavesdropping in her early field notes reflected the dissonance between her insider status as an islander and her outsider status as a social scientist. "I noticed Gwen, Dick, Nobbs family switching to modified Norfolk when speaking to dad," she wrote in 1970, "was this because I was present, or a common feature of their speech these days?"[84]

Harrison's most novel technique, made possible in part through the appearance of portable cassette-based tape recorders, was to enlist locals as fieldworkers. She handed out blank tapes, with loose instructions: "It would be excellent if you could make a recording in three separate conversations of about 20 minutes each. The subjects that are talked about are not terribly important as long as they are topics that are easily discussed in Norfolk."[85] As Harrison's conscripted fieldworkers roamed the island, recorders in hand, they elicited improvised encounters and conversations of their own. When one volunteer's machine attracted the interest of a four-year-old child, the record of two adults coaxing "Broad" Norfolk speech out of her proved a valuable record in its own right.[86] Harrison's informants were not guinea pigs in a laboratory, but active agents in the preservation and study of their own languages, with substantial freedom to produce their own dialogues on their own terms.

After finishing her interviews on Norfolk, Harrison returned to Sydney, where she continued to record members of the Norfolk Island diaspora and set to work producing a dissertation, "Variation in Present Norfolk Island Speech," which she defended in 1984. In that text, and in her subsequent professional publications and correspondence with fellow academics, she entered into debates about the status and categorization of Norfolk. Like her fellow linguists, she increasingly read the language itself against the island's history of colonialism and scrutiny—but she departed from her colleagues by suggesting that Norfolk was best understood not in diametric opposition to English, but as on a continuum with it. She made an argument about code-switching, and did so as a consummate code-switcher herself, not only as a woman practiced in switching between Norfolk and Australian English, but as a Norfolk Islander speaking in the languages of academic linguistics. Harrison suggested that her dual status as a Norfolk Islander and a field scientist made for a generative recording praxis, establishing an "INSIDER-OUTSIDER role in interviews with young people. Recognition of her INSIDE links with the Norfolk Island community encouraged informants to interact normally with the interviewer and with one another. At the same time, she was able to act as an OUTSIDE observer, a friendly supervisor and guide to the progress of each interview."[87]

Most of her 450-page dissertation consisted of excerpts from her field recordings, which she accompanied with close readings to produce an analysis of the island's variegated linguistic topography. She broke her informants down into five distinct categories: adults who spoke an almost pure "Broad" Norfolk, adults who were competent "Broad" speakers, adults who spoke "modified" Norfolk, youths who spoke with many "Broad forms," and youths who used many "modified forms." Both sets of youth also employed forms that were unique to their generation. All of her informants engaged in code-switching between some variety of English and some variety of Norfolk. Speakers often chose from a linguistic repertoire to signify their identities; if an islander was asked whether they spoke Norfolk, "an answer in the affirmative was the speaker's ticket-of-leave into a network of Norfolk Island camaraderie, allowing him to be accepted as a 'true' Norfolk Islander sharing insider status and obligation and exchange rights with others of the same racial background," Harrison wrote.[88] Moreover, those shifts happened fluidly. By selecting some words or sounds over others, a speaker could shift, often within a sentence, between identities and meanings. As Harrison put it: "Having Broad Norfolk and Norfolk English as symbols of WE/THEY, INSIDER/OUTSIDER association enables the Norfolk Islander to exploit the stylistic potential of the intermediate area between codes. Thus a move towards the THEY code may coincide with change of topic or participants in the conversation. It may signal information about the speaker's feelings and attitude towards his subject, towards his interlocutor and towards his language setting."[89] Perhaps most importantly, Harrison suggested that while shifts in code were often intentional and situational, they also became integrated into the unconscious structure of the language itself, evolving as a permanent feature of speech on Norfolk Island. It was all perhaps best understood as "style," whether deployed intentionally or otherwise.[90]

In a 1986 paper, she suggested that these changes could happen rapidly and fluidly. Below is a dialogue she recorded between a twelve-year-old child (ChM) and an older informant (TM) at the island school; Harrison underlined its "modified forms" for emphasis:

> ChM: Should *bi sə si ə fʌs* time *wi bɪn* camping was *dʌun hj ʌ had wʌn* big line between *I tu tri soə wɪ* thought . . . well, *wɪ nœwə gʌt wʌn* tent *soə wɪ tʃʌk wʌn* bit *ə* canvas *oəvə* the top of *da* wire so the tent's in a pyramid without any sides on it *ən* I went to sleep about nine o'clock *ən* E. *ən dɛm stɛ əwek ən* I woke up about three o'clock *ən lʊk* in the end of the tent *ɛs* this *goəsbird hɛm ən mi dɛm sə* already *stat ʌp kas dɛm sə* scared. Him *ən* G. *ən I* dɛə . . . ooh, I *nœwə bɪn flaɪ soə fas*
>
> TM: I've *nœwə* seen you fly.
>
> ChM: *ju nœwə bɪn si* anything *dɛn.*

Harrison also provided a version in Australian English:

> ChM: You should have seen the first time we were camping. It was down here, and there was a big line between these two trees, so we thought. . . . Well, we didn't have a tent so we threw a bit of canvas over the top of the wide . . . and E. and they stayed awake and I woke up about three o'clock and looked. In the end of the tent was this ghostbird . . . him and me . . . they had already gone because they were so scared . . . him and G. and I were there . . . oh, I have never flown so fast.
>
> TM: I've never seen you fly.
>
> ChM: You've never seen anything then.[91]

In a sense, Harrison argued that Flint and Ross had perhaps asked the wrong questions. It was difficult for Flint to determine a clear line between English and Norfolk because no such clear line existed. It was difficult for Ross to locate pristine histories and etymologies in practice because these spaces were never pristine, and situated adjustments to the presence of strangers were long-standing features of Norfolk's language. The idea of hard barriers between the island and the wider world, and notions that it was a "pristine" laboratory rather than a living space, had deranged their vision.

By the 1980s, academic linguists increasingly understood Harrison as a crucial source. Scholars compiling catalogs of contact languages or English varieties relied on her when building their entries for Pitcairn and Norfolk. The eminent creolist John Reinecke based the "Norfolk" entry in his 1975 *Bibliography of Pidgin and Creole Languages* in part on her master's thesis, calling the island "a laboratory case in creole formation.[92] Linguist John Holm, too, labeled the language a "creole" in large measure on the basis of Harrison's work.[93] Academics saw Harrison as a vital go-between who could relay knowledge from an insular place. She supplied Holm with interview transcripts and translations, which he published in his own volume on contact languages.[94] She translated fifty lines from English into "Broad" Norfolk for University of Texas linguist Iain Hancock's comparative project on creole languages.[95] University of Heidelberg linguist Manfred Gorlach, who had read some of Harrison's previous work, wrote to her about potential publication in a series he was editing. However, perhaps mirroring Rosalind Young's treatment by later historians, Harrison's fellow linguists too often saw her as a source of data rather than theory. Gorlach, for his part, inquired about her "access to the in-group of elderly speakers of N" and asked her to better define her own skills with the language.[96]

Even so, Harrison worked to defend the legitimacy of her interpretations and, concomitantly, the status of the language as a separate language in its own right. One of her first debates was with Swedish linguists Arne Zettersten and Anders Källgård. Zettersten had conducted fieldwork on Norfolk in 1970.[97] Källgård, a student from the University of Gothenburg, went to Pitcairn to make recordings of his own in 1980.[98] When Zettersten sought to jointly publish some of their results in Manfred Gorlach's *English World-Wide* series, Gorlach encouraged the Swedes to correspond with Harrison, especially after noting that their research portrayed the islands' languages as less distinct from English than she and Flint had.[99] Harrison offered an explanation, founded on her different methods of fieldwork. "No doubt the interview situation," particularly the fact that English-speakers and "Broad" Norfolk speakers were conversing together, was an impediment.[100] She also critiqued their translations and orthography, and especially Källgård's choice of English spellings over IPA transcriptions. Källgård sought a standard spelling across their publications, suggesting that they adopt English standards where possible, limit IPA transcriptions, and provide few translations. Harrison fervently disagreed, cognizant of the political and cultural stakes: "I feel that it misrepresents the character of the language," she wrote, suggesting instead that any standardized system "should indicate as clearly as possible the characteristics which distinguish Norfolk from English." An English-based orthography would dull the uniqueness of Norfolk and flatten the variations Harrison had identified. Conversely, IPA would capture differences of pronunciation, and translations would by definition assert Norfolk's cultural autonomy.

A more heated and long-running set of debates emerged with Donald Laycock, a linguist at the Australian National University who had written prolifically about contact languages—indeed, Flint had based his Norfolk intelligibility tests in part on Laycock's field methods in New Guinea. During the 1980s, Laycock became interested in Norfolk after he was invited to help author a book on its language. Alice Buffett, a Norfolk Islander and cousin of Shirley Harrison, won a grant from the Churchill Trust to design a system of writing for Norfolk, part of a general turn toward the preservation of threatened languages. The award sent her to ANU to take a course in linguistics and provided money to hire an expert consultant. Laycock was initially hesitant to take the job; he told Harrison that he had recommended her as the better candidate, and in any case he was more interested in saving an endangered language than in trespassing onto her territory. He sensed that, because the invitation came from Alice Buffett and not from the island government, he was about to enter into a contentious position and he certainly did "not wish to get embroiled in island politics."[101]

Laycock nonetheless took up the work, spending a month on Norfolk Island in October of 1986. The ultimate goal, he explained, was a system of writing that anyone who already read standard English could apprehend, but that would appear "sufficiently different from Australian English to be readily recognized as Norfolk."[102] To achieve that system, Laycock spent much of his time putting Harrison's informants, especially her father's interviews from her 1972 thesis, in dialogue with his own. Laycock claimed to produce a workable system "by looking at [Harrison's] sentences, then composing sentences of the same type (taking wild guesses that they might be acceptable in Norfolk), then checking them with Alice."[103] Dialogues were in dialogue with dialogues, as fieldwork in the present called forth the resonant echoes of fieldwork in the past.

Laycock's work also re-amplified long-running disagreements between Norfolk Island families, and between Harrison and her fellow linguists, some of whom were already committed to different systems for writing the language. The adoption of one orthography or another carried implicit stakes for the status of the language itself. Some islanders had already published literary writing using one form or another; whose family's forms would become standard? And, ultimately, how separate from English would the language appear? Those questions were no mere abstractions; while Laycock and Alice Buffett were working on their book, another islander, Beryl Nobbs Palmer, produced *A Dictionary of Norfolk Words and Uses*.[104] Laycock didn't think much of it; he sent Harrison a scathing multipage review of the text and said it was "so bad it gives me physical pain every time I look at it." Harrison, for her part, chided Laycock for becoming too personal, and for losing his sense of social scientific objectivity in the island's tangle of personal politics. After reading his bitter review, she wrote that "I accept the content of what you say about Beryl [Nobbs's] publication, but would like to avoid the very divisive tone that you adopt throughout the article." However, like any researcher, Harrison herself was hardly immune to politics. She tended to communicate with Laycock rather than Alice Buffett, in part because she felt "disappointment at (certain) Norfolk Islanders' ~~reaction~~ response/recognition to my own work on Norfolk Language." She was glad to see others write about the language, but she wanted her successors to recognize all that she had done for Norfolk, and to "act in an ethical manner by acknowledging the ~~sound~~ extensive and I hope thorough analysis that has already been done on the language."[105]

In addition to coauthoring his book with Buffett, Laycock produced his own academic work on the island, revisiting the contentious issue of how its language should be categorized. He sent a draft of an article to Harrison for comment in 1987.[106] In it, like many writers before him, he pored over the history of visitors' interactions with the island. Pitcairn's tumultuous past fascinated him; he wrote that "it is probable that no language-variety in the world has a more romantic history."[107] Drawing especially on Maude's historical work, and primary source accounts from Folger, Beechey, and the larger corpus of early Pitcairn texts, he noted that early visitors were almost unanimous in describing the language of the island as English—and descriptions of Tahitian-English creole were essentially absent from the record. He examined the language's form, content, and structure, too. Its grammar and syntax were essentially English, he decided. Its lexicon consisted mostly of words with English etymologies. "The lexical deficiencies of [Pitcairn-Norfolk] are such, in fact, that the language can hardly be said to have advanced much beyond the jargon stage," he declared. "The many texts given by Harrison (1984) show how difficult it is for even fluent speakers of [Pitcairn-Norfolk] to carry on a conversation without the use of words that the speakers acknowledge to be 'English.'"[108] Instead, he suggested, the language was not a formal "language" at all but a kind of "code" or "jargon" whose purpose was to separate insiders from outsiders. It was, he said, "a deliberate creation of adults, a secret language, designed to exclude non-Islanders. The proper technical term for such a language is not dialect, or creole, but *cant*."[109]

In Laycock's description, a long history of anxieties over Pitcairn and Norfolk's scrutability and an awareness of its iterative and problematic laboratory of scientific scrutiny and colonial power all cohered. What linguists studied on Pitcairn and Norfolk, he seemed to suggest, wasn't language but an anti-language, a system deliberately conjured to make the discovery of true knowledge by outsiders impossible. It was, in multiple senses, a kind of de-ranged vision, reliant as it was on both a notion of profound separateness and on paranoid visions of deliberate, intergenerational obfuscation and duplicity. Harrison disagreed vehemently with Laycock and wrote back to dispute his vision. She, too, had read the same historical sources, and argued from them that outsider contact with the island during its first decades was far too infrequent to merit the construction of a private jargon.[110] Moreover, while she was the first to argue that expressions of Norfolk were fluid, giving rise to the "modified" forms she had spent a half decade recording, "Broad" Norfolk had remained remarkably consistent over the past six generations—it was, in her words, "an exceptionally stable language which has maintained as many T[ahitian] and English dialect features from the time of its origin."[111]

She implied that Laycock had demeaned Norfolk. "Your approach almost seems to give support to the old view that pidgins and creoles are such simple, inferior forms of language that they are not worth much attention," she accused.[112] Their views could not have been more different, and offered competing visions of both the language and the island community that spoke it. Harrison saw a creative continuum of linguistic modes, from a stable Broad Norfolk, to modified forms, to a Norfolk Island English, all of which had emerged as important markers of community, history, and tradition. Laycock, echoing Moverley's diatribe decades earlier, saw a language of deliberate obfuscation, meant principally to thwart prying outsiders like himself. It was an almost too perfect manifestation of two centuries of investigative scrutiny, and of epistemological anxiety that island laboratories were anything but laboratory-like.

ENTANGLED TAPES, REVERBERATING HISTORIES

Laycock died unexpectedly in 1988 at the age of fifty-two.[113] His articles on Norfolk were published posthumously with the help of his former ANU student Peter Mühlhäusler, who was became an authority on contact languages in his own right.[114] In the next decades, Mühlhäusler, too, was drawn to Norfolk Island as a field site. He had first encountered the Pitcairn and Norfolk languages while reading Ross and Moverley's 1964 book and became fascinated.[115] From 1997 on, he visited the island nearly every year, authoring dozens of papers on the language and its history.[116] He supervised a graduate student, Joshua Nash, who studied Norfolk Island place-names and who went on to do the same on Pitcairn.[117] Like Laycock, Mühlhäusler collaborated with Norfolk Islanders to write about and document their language and culture.[118] "People have mixed views about Peter [Mühlhäusler]," Norfolk Islander and author Piria Coleman told me. "He has been incredibly generous at the same time as expecting a lot from local people. It wouldn't matter if he knows I've said that. The fact he has made the effort to come back for so long has not gone unnoticed and that is the way to earn respect in a small place like Norfolk."[119]

During the decades surrounding the turn of the millennium, the steady interest of linguists in their language helped some Norfolk Islanders very directly in their bids to secure its categorization as a language in its own right. In 2004, the island's legislature passed a law to recognize Norf'k—Mühlhäusler's and some islanders' preferred spelling—as an official language. In his argument, the law's sponsor recounted the history of the language and its recognition by academics: "It has experienced testing times. It has been scorned. It has been scoffed at and it's been derided. It has been banned in the Norfolk Island educational system. Physical and mental punishment has been meted out to those who have entertained its use but notwithstanding all of this, the Norfolk Islander has not been dispossessed of this essential cultural trait. Indeed in more recent times we have turned to teaching it in our school and it has been the subject of a number of learned treatises."[120] In the years since, Norfolk Islanders have worked on language-planning programs and advocated for its preservation.[121] The categorization of the language itself, however, remains subject to debate in linguistic circles; some argue that it is a creole, while others are agnostic on the question.[122] However, despite Flint's and Harrison's extensive work to overcome hesitancy in the field, and despite Moverley's and Laycock's anxious worries over deliberate obfuscation, the notion that Norfolk and Pitcairn are natural laboratories thrives, too.

Writing in 2008, Mühlhäusler and coauthor John Ingram reiterated that both Pitcairn and Norfolk continue to "provide laboratory conditions to study linguistics processes such as language contact, dialect mixing, and languages in competition."[123] As in the case of so many other disciplines, the metaphor endures.

Tape recordings and transcripts, the records and products of that uncannily persistent metaphor, survived as well and continue to circulate among researchers, Pitcairn Islanders, and Norfolk Islanders into the present day. Not all survived, of course; Moverley's 1951 tape disappeared sometime after Ross's death in 1980, though linguists often asked each other if it could be tracked down.[124] Flint's tapes, on the other hand, traveled widely. A copy went to a linguist in Scotland, in exchange for tape recordings of a Tristan da Cunha speech.[125] Another copy went to linguist Ralph Gardner White in Tahiti, in exchange for help with Tahitian etymologies.[126] Flint was especially generous about sharing his tapes with Shirley Harrison; he gave her everything he had ever recorded, alongside copies of tapes from Harré and Gathercole.[127] When he died in 1984, he left her his library of sociolinguistics books.[128] Pitcairn and Norfolk dialogues, rerecorded, retranscribed, and often digitized, circled back to the sites at which they were born. Versions of Flint's dialogues are on display in the Norfolk Island Museum, and are archived on Norfolk Islanders' hard drives.

These documents are not perfect records, having become warped with time and tangled in their journeys across and beyond the Pacific. Materially, tape was prone to break with storage and use. Harrison snapped Gathercole's tapes in three places, though she stitched them back together as best she could, which resulted in only a few losses.[129] Ross and Grimson wrote that Moverley's tape was so worn with play as to be incomprehensible. Time and distance, as well as frequent handling, left ineradicable marks and intractable entanglements. Even today, as cultural and linguistic archives, tape recordings remain entangled in long and complex histories of iterative surveillance both by scientists and by the colonial state.[130] They hum with noise and static, literal and figurative, especially as one field experience informed another and recordings begat more recordings. With each new field encounter, field methods and epistemological anxieties were not just reproduced but re-amplified. Despite pursuing notions that Pitcairn and Norfolk were "pristine" "laboratory" spaces, fieldwork in spaces of iterative scrutiny demanded the elaboration of new and collaborative methods—which in turn evolved into ever more agentive forms as Flint's interviews became Harrison's. At the same time, debates over the categorization of the language itself, of a piece with larger debates over the cultural and political autonomy, grew from arguments over the boundaries between dialect and language to assertions that the Pitcairn and Norfolk languages were in fact secret codes, designed to be inscrutable to those strangers trying to listen. In Moverley's and Laycock's analyses, the reverberations of two centuries of scrutiny and surveillance, as well as resonant anxieties over the limited capacity of these supposed laboratories to produce easy knowledge, rang noisily, almost deafeningly.

And so, before closing, let us take Flint's reel off our tape player and listen again to Moverley's recording—only figuratively, of course, since it survives now only as a transcript in a half-century-old linguistics text, and this is, after all, just another textual reproduction. But no matter. We fast-forward and then slow the tape down. Once more we hear Moverley (M) and his former pupil (P) talking around a microphone in Wellington in 1951. They are imagining themselves on Pitcairn, where they see some young people clumsily carry a baby:

34. M: Now that they know that we are watching them, the children are behaving better.
    P: iˋjeə | demʑ ˈʔaɪ sə ˋpiəɫ | *They certainly are; they're feeling ashamed.*
35. M: Oh, they are ashamed, are they?
    P: iˋjeə | kɔːwʌn ˋstreɪndʒə ˌsi dem | *Yes, indeed, because a non-Pitcairner is watching them.*[131]

Conclusion

Unmaking the Natural Laboratory

The myth of Pitcairn and Norfolk Islands as human laboratories has lived on well into this century. Both communities remain as studied as ever, attracting researchers interested in deploying them as scientifically or intellectually generative microcosms. Like the anthropologists, linguists, historians, and others who preceded them, scholars today continue to invoke the islands' romanticized histories in order to construe them as uniquely isolated and legible. Take, for instance, this portion of the abstract of a 2015 paper appearing in the journal *Investigative Genetics,* which very neatly folds together the mythology of the *Bounty* mutiny and an image of insular legibility: "The Pacific Oceania region was one of the last regions of the world to be settled via human migration. Here we outline a settlement of this region that has given rise to a uniquely admixed population. The current Norfolk Island population has arisen from a small number of founders with mixed Caucasian and Polynesian ancestry, descendants of a famous historical event. The 'Mutiny on the Bounty' has been told in history books, songs and the big screen, but recently this story can be portrayed through comprehensive molecular genetics."[1] The long-standing idea of the Pacific as regional laboratory, defined by its recent settlement, is married to the romantic history of the *Bounty* mutiny, all in order to configure the Pitcairn-descended community on Norfolk Island as uniquely useful to science on essentially racialized grounds.

The above-mentioned article was produced by a team of geneticists from Australia's Griffith University who have, in collaboration with researchers from other universities around the globe, studied Norfolk Islanders for two decades. They have pursued two basic research trajectories. First, they have sought to determine whether specific health problems—eye disease, heart disease, and migraines, particularly—are linked to specific genes.[2] Second, they have sought to substantiate Norfolk Island as a usefully "admixed" "isolate" for genetic research, work that suggests rhyming similarities with the physical anthropology of the early twentieth century.[3] Like Harry Shapiro before them, geneticists combed through written accounts and combined these with genetic analysis of "ancestral informative markers" to produce a comprehensive pedigree of 5,742 Norfolk Islanders, a web of parent-child relationships spanning centuries mapped with an expansiveness that far exceeds Shapiro's tables and note cards. The explicit language of "race" is gone now, replaced with "haplogroups" identifiable from specific genetic markers—though these retain the highly racialized categories "European" and "Polynesian."[4] A century on, the island is still seen in essentially the same ways Charles Darwin, Alfred Russel Wallace, and Harry Shapiro imagined it—a kind of isolated racial laboratory useful for making apparent the mechanisms of heredity and the inheritance of traits between disparate groups.

Invocations of Pitcairn and Norfolk Islands as human laboratories often rely on a selective vision of the past. The *Bounty* mutiny is celebrated again and again; the mutineers are central protagonists in most articles' introductory narratives, including much recent work in genetics. The islands' founding generation is described as that of the male mutineers first and their female, Pacific partners second, an order of precedence that replicates the masculine-centered fixations of the past. The move to Norfolk is invariably featured, too, a refounding of the accidental experiment that makes any given study possible. Other pasts are undertold or unspoken. Pitcairn and Norfolk's histories as sites of colonial investigation remain especially obscured; the features that make the islands useful as places of supposedly singular scientific scrutability tend to emerge as happy accidents rather than as the deliberate products of colonial experimentalization. Across the disciplines—from geneticists' "isolates," linguists' "pristine" sites, to anthropologists' "laboratories"—the deployment of Pitcairn and Norfolk as research sites is often reliant on selective rememberings and forgettings, and especially the elision of the colonial interventions and configurations that made the two communities laboratory-like in the first place.

Forgotten, too, is what the tremendous and ongoing accumulation of these scrutinizing investigations has meant for the people subject to them. "It feels like living inside a petri dish sometimes," Pitcairn Islander Melva Evans told me during a conversation near the end of my stay there, when I asked about the persistent presence of strangers like myself. Others, whom I promised not to quote in print, echoed the sentiment. "Writers," those who come to the island to extract stories and knowledge, have not left a positive impression—and there have been many. Across more than two centuries, hundreds of outsiders have visited and used Pitcairn and Norfolk Islands to advance their own ideas about the world, to produce or to question their theories, just as I have here. The islands have been and remain sites for the construction of "Britishness," for projecting imperial ambition, for questioning the nature of biological race, for understanding the parameters of language, and, in the case of the book you are reading now, for critiquing the history of academic research and its colonial entanglements. But for those subject to these investigations, it has meant unending scrutiny with deleterious epistemic and political consequences. Academe and empire, knowledge and power, were and remain thoroughly threaded together on both islands.

And so, though this last section of my book is a "conclusion," I am unsure that any conclusion to the history I have told is within easy reach. Outsiders have spent two centuries thinking with the natural laboratory; unthinking it will be hard. Rather, to bring my own narrative and my own meddling use of Pitcairn and Norfolk Islands to an end, I will ruminate briefly on the ways the iterative and persistent scrutiny of outsiders has not only continued to shape the writing about Pitcairn Islanders, but also remained imbricated with colonial interventions into their lives. Then, I will gesture toward some ways a cognizance of that ongoing history can open up alternative scholarly visions, not just of Pitcairn and Norfolk Islands, but of the Pacific and perhaps even of colonized and Indigenous subjects generally—with a hope that the myth of the island as natural human laboratory can finally be dismantled and consigned forever to the past.

OPERATION UNIQUE AND COLONIAL PERDURANCE

One of the most vexing cases in which histories of iterative investigation continued to shape Pitcairn Islanders' lives in the present was a series of criminal investigations and trials conducted during the first decade of the new millennium. In 1999, a British police officer visiting Pitcairn Island to train its local government discovered serious and credible allegations that several men had sexually abused the community's children. The resulting criminal investigation, Operation Unique, saw two police officers from the Kent constabulary interview members of the Pitcairn diaspora in Australia, New Zealand, and Norfolk Island. They identified thirty-one victims and twenty offenders, whom they suspected of rape, indecent assault, and incest in crimes that dated as far back as 1960.[5] Operation Unique, whose name itself gestured at the ways the island was understood as exceptional, made headlines around the world, and saw a reapplication of the same discourse of utopia and dystopia that had long inflected accounts of Pitcairn. Journalists, deploying the same language of disillusionment as their predecessors in centuries past, suggested that their investigations cleared away layers of mythologized artifice to reveal "reality." As one writer for the *New Zealand Herald* put it: "The allegations could see Pitcairn Island relegated from its fanciful position as an island paradise founded on contempt for mindless British authority, to a vestige of male chauvinism and brutality."[6]

This episode has been much written about, and often not in ways that are particularly respectful to the community itself, which once more was disparaged in exoticizing accounts before international audiences. I revisit the episode here not in order to breathe new life into these narratives, but rather to investigate the ways the affair was itself understood and reframed by outsiders for their own purposes—and especially to make visible the ways that long histories of iterative investigation and representation encrusted legal and journalistic visions of the island.[7] The criminal investigations and resulting trials became an occasion for journalists, academics, and the remnants of the British colonial state to once again wrestle with questions of identity and colonial rule on Pitcairn Island. Like their predecessors, whose records often formed the basis of their own investigations, they found that the island fundamentally deranged their assumptions about the nature of British colonial sovereignty not only in the past but in the present—even as they also contributed to the maintenance of imperial rule.

Even before any case went to trial, it became clear that the British state would reactivate the machinery of empire to intervene dramatically on Pitcairn Island. Across the preceding decades, British officials had increasingly come to view it as a burdensome relic of a bygone era. One administrator, writing in 1979, declared that he "would dearly like to wipe our colonial slate clean in the Pacific" and lamented that not all his colleagues shared that view. "Some of them," he said, "even have a romantic nostalgia for such anachronisms as Pitcairn and would strongly oppose any drastic action to get shot of it."[8] But Pitcairn, like Norfolk, was never postcolonial. It remains a formally colonized space, and the reason, in the most literal sense, that the sun does not set on the British Empire. While the criminal investigations proceeded, some figures in the Foreign and Commonwealth Office suggested a general amnesty and a truth and reconciliation commission in order to address the offenses without destroying the community.[9] Others offered much more severe prescriptions, wondering "whether or not the community on Pitcairn should continue to exist," referencing the possibility of forced migration with a casualness well beyond Henry Evans Maude's proposed relocation schemes a half century earlier.[10] Matters were finally decided when Baroness Patricia Scotland, the minister responsible for Britain's remaining overseas territories, ordered the case brought to trial.[11] As a result, defendants on Pitcairn itself would be tried, via remote link, by judges in New Zealand, under the architecture of British imperial law.

Following the decision, the Pitcairn Island government, led by Mayor Steve Christian, asserted its right to decolonize. In a presentation at the Pacific Regional Seminar, Christian argued that the trials threatened the island's way of life, and "that the Pitcairn people will no longer be able to exist in our homeland."[12] Another islander, Kevin Young, spoke at the Caribbean Regional Seminar, offering a "Pitcairn Decolonization Statement."[13] The United Nations formally requested that the UK government "keep the Secretary-General informed of the wishes and aspirations of the people regarding their future political status."[14] The Foreign and Commonwealth Office responded by asserting that it had renewed its interest in and funding of Britain's overseas territories—it had become clear that continuing with the prosecutions would also mean rearticulating Britain's legitimacy to rule. Indeed, questions of sovereignty and the right of the empire to rule would become core issues in the legal proceedings to come.

Hearings began in 2004. Ultimately, six men were convicted of thirty-five offenses and sentenced to imprisonment in facilities on Pitcairn, which they had built themselves while awaiting trial. These verdicts were immediately appealed, a process that would last until 2006, when the case was heard by the Judicial Committee of the Privy Council, the supreme court of appeal for the last vestiges of the British Empire. Throughout, both prosecution and defense operationalized the island's long history as an object of study in order to build their cases, and breathed new life into old questions about its supposed exceptionalism. The defense sought to challenge Britain's right to try the islanders in the first place, arguing that Pitcairn Island was founded in mutinous rejection of imperial authority, that many of its founders were not themselves British, and that the colonial government had only intermittently and superficially intervened in its governance. The public defender produced a "lengthy chronology" of the island, which was accompanied by nine volumes of supporting material, totaling thousands of pages and stretching back to the early nineteenth century. Figures like John Barrow, Rosalind Young, Harry Shapiro, and especially Henry Evans Maude made up much of that archive.[15] The trials thus not only replicated earlier colonial investigations but were built on top of and from them. Ultimately, however, it was only imperial fiat that mattered. The prosecution argued that Britain's legal authority stemmed entirely from the British Settlement Act of 1887, which Colonial Secretary Joseph Chamberlain had extended to Pitcairn Island by administrative declaration in 1893.[16] The Privy Council agreed. In his judgment, Lord Hoffman wrote that it was "unthinkable" that the Judicial Committee would undo the Crown's colonial sovereignty, adding that the "reality of British governance is not shaken by the argument that the legitimacy of its establishment is flawed."[17] All appeals were dismissed, and the convictions would stand.

Decisions of the Judicial Committee of the Privy Council rarely featured dissents, but in this case, Lord Hope offered an alternative opinion. He agreed that the appeal should be denied and convictions sustained, but insisted that the court engage more substantively with the historical and philosophical underpinnings of Britain's colonial legitimacy to rule Pitcairn—that the island, in a sense, made abstract questions of imperial sovereignty "thinkable."[18] Indeed, like so many practitioners of other disciplines, he thought that this case, far from providing easy answers, raised "fundamental issues about the rule of law" and touched on core principles of governance; one of his citations was Hobbes's *Leviathan*.[19] His opinion, which has since been augmented by several legal scholars, revealed in stark terms Pitcairn's power to derange the assumptions and received notions of those who studied it, even in matters of jurisprudence and political philosophy.[20] In that way, Hope's opinion, which wrestled with the fundamental basis of the rule of law itself, was in keeping with the long history of knowledge making on Pitcairn, which so often saw investigations of Pitcairn and Norfolk Islands prompt derangements of core disciplinary categories.

Deranging visions emerged in other quarters, too. Unlike Hope's colleagues on the Privy Council, journalists reporting on the Pitcairn prosecutions were more eager to interrogate fundamental questions and to locate in the criminal case broad "lessons" for the wider world. Among them was Kathy Marks, who covered the trials in person and eventually published a widely read book on them. Her account joined a long, paranoiac discourse that had always imagined dystopian secrets lurking beneath the island's supposed hospitable facade. Her lurid title made that perspective clear: *Lost Paradise: From Mutiny on the Bounty to a Modern-Day Legacy of Sexual Mayhem, the Dark Secrets of Pitcairn Island Revealed*.[21] Marks read many of the accounts produced about the island over the past two centuries; I occasionally found her yellow request slips falling from folders in the

British National Archives while conducting my own research. Echoing a long undercurrent of epistemological anxiety about imagined local artifice, she understood the island's history of engagement with outsiders in entirely pragmatic terms, writing: "If the world learned what Pitcairn was really like, the locals feared, the cruise ships would stop calling, and the parcels of new cloths would stop arriving . . . the myth would be exposed for what it was, just a myth, and the mutineers' descendants would be left to their own devices."[22] Like many others before her, she suggested that the islanders' efforts to "hypocrite the stranger" were nothing more than a cover-up, a means by which they could blind their guests to the *real* Pitcairn Island lurking out of sight. It was a move not dissimilar from that of anthropologists like Shapiro a century earlier—to reveal "dark secrets" others had missed, rendered her telling more interesting, more singular, more valuable.

At the same, Marks's vision of the island as cloaked in artifice was accompanied by a parallel assertion that it remained a space of microcosmic legibility, with obvious lessons for the wider world. Comparing it to William Golding's *Lord of the Flies,* she wrote that "like his unnamed island, it was a social laboratory, but a real one—the site of a unique experiment thrown up by the confluence of historical events."[23] And, as a laboratory or experiment, the island for her seemed to gesture at very large and unsettling truths: "What did Pitcairn tell us about human nature and life in small, remote communities?" she asked. "Is this how all of us would behave if left to ourselves, with no one looking over our shoulder? Is Pitcairn a cautionary tale?"[24] Indeed, she built toward a broad theory of violence and masculinity applicable well beyond the island. In her account, the island was deranged from the beginning, a place where men went mad with patriarchal power. Mutineers came to an isolated island and, after a string of murders of women, built a patriarchal society that treated them as sexual commodities.

However, if we read the trials' juridical and journalistic accounts against a wider history of investigation and colonial interference, I suggest that we might learn as much about the interests, anxieties, and myopias of investigators as we do about the investigated. For one thing, we might learn that the language of disillusionment is very old, and that long before Marks, outside observers had promised to reveal "realities" lurking beneath utopian facades. For another, the eugenic investigations historicized in chapters 2 and 3 show us that colonial attempts to discipline the island in "moral" terms have a very long history, and that investigations into those moral concerns were accompanied by and justified much wider colonial infringements of cultural and political autonomy.

Observers often tended to fixate on matters of sex and reproduction in their accounts, in no small part because racialized understandings of "heritage" and "inheritance" from the founding generation to the present had always remained the core locus of their interest. There was a reason anthropologist Harry Shapiro titled his book *The Heritage of the Bounty*.[25]

Nearly all investigations about Pitcairn Islanders centered on "legitimacy" in some form, often linking the disciplined management of heredity with the right to political autonomy. Houston's investigations on Norfolk island focused on "illegitimate" births in order to assess the community's success as a colony (chapter 2). Historian and colonial governor Harry Maude's archive (chapter 4), so well represented in court documents, showed the same; he investigated the islanders' parentage, collecting a list of "illegitimacies" as part of his official inquiries.[26] But this was true of academics as well. Shapiro's genealogies, haunted by what he called the specter of "sub-rosa promiscuity," called into question the viability of racial mechanics as an object of study and even race as a category (chapter 3). More abstractly, even linguists like Laycock became obsessed with the notion that the islanders' language was not a proper creole, the legitimate fusion of two parent languages, but almost an "illegitimate" offspring designed to obscure meanings from outsiders (chapter 5). In studies of Pitcairn, everything rested on understandings of heredity and reproduction as proper or improper, legible or secret. In that way, Pitcairn's archive reveals how the disciplining of reproduction remained intertwined with visions of colonial order well into the twentieth century, a persisting manifestation of what Ann Laura Stoler has called "a racialized common sense about people and places."[27] Even in the case of investigations into credible accounts of sexual violence like Operation Unique, the nature of the ensuing prosecutions, their jurisprudential reception, and especially their recounting in journalistic writing did not occur in a historical vacuum.

That long, iterative history mattered. In recounting the British colonial state's prosecutions on Pitcairn in this century or the Australian state's moralizing colonial scrutiny in the last, I aim not to resurrect difficult moments in order to make any claims about the islands themselves. Rather, I suggest that outsiders have long scrutinized Pitcairn Islanders for their own purposes, and that these investigations were often bound up with colonial rule. Moreover, I also suggest that those who studied and governed Pitcairn and Norfolk were members of insular communities in their own right, with their own insular derangements. In places as consistently and iteratively studied as these two islands, investigations piled up, with consequences for the shape of the discourse and investigations to come, as well as for the communities subject to them. Visitors encountered the lasting reverberations of their predecessors' visits and often became obsessed with the haunting possibility that, even in places celebrated for their scrutability, their attempts at locating truths might be somehow occluded. Indeed, I close this book with Pitcairn's trials neither to join Kathy Marks in suggesting that the islands are actually hidden dystopias, nor to hint that there is a "real" Pitcairn or Norfolk lurking somewhere behind layers of artifice. Rather, I take both notions as recurring products of a long history of outsider fascination—a perduring interest that, in the case of the island in the first years of this century, resulted in considerable colonial intervention, the reanimation of two centuries of investigative reportage, and the recrudescence of very old myths of utopian or dystopian exception.

THE VIEW BEYOND THE CLIFF'S EDGE

But what else is possible? What other visions can we call into being? The idea of Pitcairn and Norfolk Islands as human laboratories was two centuries in the making; to unmake it will not be easy. One common starting point for deconstructing lingering colonial structures is a disciplinary reflexivity, an awareness of what has come before and the ways scholars' efforts to produce knowledge remain imbricated with efforts to produce power. Certainly, I have tried to show how that was the case on Pitcairn and Norfolk. In the preceding chapters, diverse sets of knowledge makers—sailors, colonial administrators, anthropologists, historians, linguists, and others—leveraged the islands for their own purposes and to construct their own "truths," even as they contributed to both communities' broader perdurance as exemplary spaces and natural laboratories. There is, however, another recent knowledge maker who has studied both Pitcairn and Norfolk Islands to produce his own grand theories, and whose work I have not sufficiently historicized: me.

I first encountered Pitcairn and Norfolk Islands in the same way Harry Shapiro did, as examples of a "natural experiment" in human genetics, during an undergraduate physical anthropology course at the Ohio State University in 2005. Later, as a doctoral student in the history of science at Princeton, I chose to study the studiers of Pitcairn Islanders in part because I remembered that example, and wanted to unpack the ways researchers and their human subjects interacted in the creation of knowledge. However, I also chose the subject because I had read with fascination histories by Henry Evens Maude and especially his protégé, Greg Dening. I soon consulted archives in the US, UK, Australia, and New Zealand, and studied nearly every published text about both islands. I visited Norfolk Island for ten days in May of 2014, and Pitcairn for eighteen days in February and March of 2015.

Like so many other researchers, my experience was touristic. I spent some of my time on both Pitcairn and Norfolk seeing sites about which I had only read, retracing the steps of those who came before me. As a trained historian, I feel most at home in the archive, where I read the filed-away papers of those who have long been dead. But on the islands, I tried my hand at a simple kind of fieldwork. I recorded a few informal conversations on Norfolk Island, though on Pitcairn I kept my camera, digital audio recorder, and notebook in my pack, perhaps too aware and too nervous about some Pitcairners' dislike for writers—though I was always very clear about my reason for being there. I jotted down my experiences in journals at night, summing up each day. I still keep those field diaries, one a digital text, the other a thin moleskin notebook with entries and illustrations in pen. Reading them now as I would any other archival document, I see the uncertain work of a novice graduate student. My past self worried a great deal about the reflexivity and recursion that inevitably arose when writing about people who wrote about people. I also see that my reflexive concern was positioned against a persisting desire to "extract" knowledge, to bring back something useful that could help in the writing of a book and the difficult bid to secure tenure. I understood, then, the colonial nature of such extractive forms of knowledge gathering—but an awareness of the politics and practices of field investigation did not eliminate the structures and incentives to produce certain kinds of work.

Those same diaries suggest that I found communities unsurprisingly accustomed to explaining themselves to outsiders. Most of those with whom I spoke knew well the work of the researchers who preceded me; copies of books about Pitcairn and Norfolk abounded on Pitcairn and Norfolk, and interlocutors often showed me their collections. One Norfolk Islander brought out a charred mass of pages, black around the edges, which he told me he had once pulled from the remnants of a house fire, intimating that arson was the unsavory result of a political disagreement about the island's autonomy. The title page on the top of the stack still clearly marked it as Ross and Moverley's *The Pitcairnese Language* (figure 3). It was a reminder of the many and sometimes surprising ways that the products of investigations accreted on both islands and became further encrusted with their own meanings. It was a reminder, too, that these communities are not the monoliths so many authors describe when using the language of "insiders" and "outsiders," but instead of course have their

Figure 3. A copy of Alan S. C. Ross and A. W. Moverley's *The Pitcairnese Language* on Norfolk Island, damaged in a house fire. Photograph courtesy of the author.

own complex social and political realities. Most of all, like other researchers before me, I wondered at the level to which I could gain "real knowledge," and what even constituted that sort of truth in places where "truths" of so many different kinds were offered by so many with such frequency. It is hard to avoid the notion that one must "break through," or that barriers of artifice enshroud a kernel of singular reality—especially after reading thousands of pages by predecessors of mine who understood their own visits in those terms. I did, however, gain a sense of what it felt like, in an embodied and practical way, to be hosted by such kind and hospitable people, and to feel the lingering presences of the investigations that preceded my own.

I also quickly learned that my research will not be the last. In fact, on Pitcairn, I was immediately replaced. During one of my last days on the island, I met another investigator, not an academic but an artist, who had arrived on the ship that would take me away. British photographer Rhiannon Adam had come to Pitcairn after the Royal Geographical Society and BBC Radio 4 offered her a Journey of a Lifetime Award. She stayed for three months, documenting her time mostly through camera lenses, though she also brought an audio recorder to produce a half-hour radio program and maintained a blog.[28] An entry in the latter relays her own sense of the place, its visitors, and the layers of myth and encounter that comprised her experience of it. "I had wondered why the islanders would risk my own negative experiences escaping into the world and tarnishing their image," she wrote after watching a visiting Danish film crew gather footage for a documentary. "But then I realized that the balance is always tipped in their favour, that the mythical and romantic image that has been cultivated for the last two centuries will outlast anything I could ever say or do."[29] It was an unsettling feeling, she suggested, insisting that "Pitcairn has led me to question what is real on more than one occasion."

Pitcairn emerges and recedes in Adam's account as it does when viewed from the ocean, a shrouded, cliff-bound mass rising or falling slowly from the waves. "For one last time I took in the steep cliffs, the foreboding shapes," she wrote of her departure. "The island was cloaked in low-lying cloud, as though it had closed up already and was hiding from me. The smoke and mirrors in tangible form, the mysticism closing in, shrouding it from view."[30] That Pitcairn should still seem occluded and inscrutable after thousands of visits and hundreds of texts is wholly remarkable, but Adam is not the only observer to feel that knowledge of the island—real knowledge—is perhaps impossible and unobtainable. I am reminded again of Lord Chelmsford's frustrated declaration, a hundred years earlier, that "there is no place where it is more difficult to find out the real truth than Norfolk Island."[31] It is an admission of unknowability and a projection of paranoid anxiety that sits uneasily alongside the notion of the island as a natural laboratory or exemplary model system, a place where isolation should permit clarity, and where knowledge is eminently legible and exportable.

And yet, I argue, those seemingly disparate visions were always intertwined—and if histories of science look deeper, their imbrications become more visible, as do the colonial structures that shaped them. In tracking their interconnected trajectories over the course of two centuries, I have suggested in the preceding chapters that persistent sites of research, not least sites of "natural experiments" or "laboratories" such as Pitcairn and Norfolk are often the furthest thing from "natural," in many instances having been carefully configured as such by a series of encounters and representations over time, first by governmental figures and then by scientists. Moreover, those who researched both islands found them decidedly un-laboratory-like in practice, often encountering frustrations that gave rise to more fundamental questions, critiques, and redefinitions that deranged the categories whose mechanics they came to study. Sailors and poets came to locate Edenic Britishness, administrators came to locate racial visions of colonial rule, anthropologists came to locate the mechanisms of racial inheritance, historians came to locate the sources of political and epistemic authority, and linguists came to locate the boundaries and definitions of languages. All came away not with easy truths, but with difficult questions and maddening reconfigurations.

I call these encounters with the limits of knowledge making "epistemic derangements," instances in which received notions or categories, especially those that drove the selection of islands as research sites in the first place, were themselves thrown into disarray. I use the term "derangement" very intentionally for its spatial dimension. The manipulation of space lies at the conceptual core of the insular natural laboratory as a notion, and derangements often resulted from collapses of the islands' supposed distance or separateness from the wider world. Pitcairn and Norfolk, like many insular spaces construed as epistemically useful microcosms, were regarded as at once *part of* and *apart from* the wider world; that was the source of their heterotopic potential. And yet, by visiting and researching these islands, investigators necessarily collapsed such distinctions. Even more crucially, researchers often became anxious that the islands were actually sites of tremendous connections and accumulations, sensing the echoes and reverberations of the many investigations and colonial interventions that preceded their own work. Far from extracting easily exportable knowledge from "pristine" sites, these visitors added another layer of entangling visits and texts on islands that had accumulated them by the thousands. At the extremes, these anxieties evolved into a manic concern that no truth on or about Pitcairn was possible, and that the island was ringed by an intense wall of artifice and subterfuge.

It is a vision that maps neatly onto physical descriptions of the island by outsiders, like Adam's fog-shrouded description of the island's "steep cliffs" and "foreboding shapes." Kathy Marks, for instance, tells us that Pitcairn's founders discovered "a natural fortress, thickly forested, with towering cliffs and no safe anchorage."[32] It is in some ways an anomalous geography. Pacific historians, especially Greg Dening, have long regarded the beach as the region's defining space, where alien cultures encountered and made sense of each other.[33] Pitcairn, crucially, has no beach to speak of, ringed as it is by rocky escarpments and pounding surf on all sides (see figure 4).[34] Nearly all accounts of the island begin with descriptions of these physical boundaries and visitors' resulting reliance on local guides to traverse them, as Pitcairners expertly rowed or motored visitors to the landing on their own launches. The mapping of imagined

Figure 4. "Pitcairn Island, inhabited by descendants of the mutineer crew of H.M.S. *Bounty.*" Engraving from the *Illustrated London News,* 1879. National Library of Australia, Rex Nan Kivell Collection, NK4182/244.

divisions on physical space continues on land: separating Bounty Bay from Adamstown is a steep Hill of Difficulty and an overlook called the Edge. These names inscribe on the island's geography a notion of profound separateness, conjuring a hard and inscrutable barrier that could be crossed only with tremendous struggle. For some visitors, those borders were insurmountable, and to locate Pitcairn's "truth" was to locate the limits of knowledge making itself somewhere in the space between the pounding surf and the cliffs above.

But such a vision of the cliff face is, I argue, the result of two centuries of deranging work by outsiders imagining the island as a place of profound separateness.[35] Writers and visitors to Pitcairn and Norfolk have long delighted in shattering myths, celebrating their own ability to recognize barriers and glimpse something real beyond the artifice, reveling in the revelation of "dark secrets" or fundamental truths. At risk of joining them, I close by suggesting that perhaps the myth of enshroudedness is itself the most enduring and indestructible myth of all, the metamyth that gives the deployments or bustings of other myths their salience. What would happen, I wonder, if scholars looked beyond the cliff face, seeing the island's more figurative boundaries with the outside world not as an obstacle to our vision but as the relic of a colonialist worldview? What if, in other words, we left the notion of the natural laboratory or natural experiment behind, ceasing to regard places like Pitcairn or Norfolk as exceptionally separate elsewheres or as epistemically freighted microcosms?

Writing on the "edges" that defined Pacific studies as well as her own intellectual career, Teresia Teaiwa—a Banaban scholar whose own community was displaced by Henry Evans Maude's forced migration schemes—invoked a Hawaiian proverb: "The top of the cliff isn't the place to look at us; come down here and learn of the big and little currents, face to face." Teaiwa suggested that the aphorism "warns against perspectives from the edge—of high cliffs—and invites a more intimate approach to knowledge."[36] Scholars of the Pacific, many of them outsiders from continental settler societies like myself, have so often written from edges, looking in, looking down. Pitcairn's imagined cliff face, its seemingly impenetrable boundary, is one of the most extreme of these edges. To imagine Pacific islands as profoundly separated and sealed off was a vital component of the myth of the natural laboratory, keeping its metonymic power alive and allowing it to represent the world but not to be of the world. But, if we look beyond the cliff face, perhaps other narratives and other visions will come into view—insular or archipelagic, local or global—and new stories might appear on the horizon.

Some of these are already fleetingly apparent, emerging as subtexts and countercurrents in the archives and on the margins. For one thing, there is the agentive power of the inhabitants of these so-called laboratories to contest the visions of those who would treat them as unspeaking objects of inquiry. Every chapter in this book has described ways Pitcairn or Norfolk Islanders intervened in and spoke back about the experimentalizing discourse to which they were subjected. Even so, agentive histories or anthropologies of Pitcairn and Norfolk Islanders remain thin on the ground. For another, so much of the writing about Pitcairn and Norfolk Islands was by men and about men, even if women's voices and perspectives, from Teehuteatuaonoa to Rosalind Young, from Shirley Harrison to the present, composed and defined so much of the islands' archival material. Indeed, obsessions with European men have been so hard for outsiders to shake that even attempts to turn scholarly attention away from the mutineers and toward what historian Robert Langdon termed "Pitcairn Island's neglected matriarchs" often recentered obsessions with masculinity. Langdon himself offered that phrase in an article arguing that Pitcairn's founding women were the descendants of sixteenth-century Spanish sailors whose ancestors had preserved their "whiteness" through "selective breeding."[37] The racialized idea of Pitcairn and Norfolk as spaces of genetic exception, harkening back to a European, male genesis is astonishingly enduring. Stories that substantively center the islands' women or that look beyond obsessions with racial "heritage" and whiteness are only beginning to be told.[38]

More generally, a view beyond the constraining boundaries of the natural laboratory might re-range Pitcairn and Norfolk Islands into Pacific history and Pacific studies.[39] No longer racialized and scientized spaces of exception, they might appear instead as constituent parts of a Sea of Islands. Without the myth of the natural laboratory, Pitcairn and Norfolk Islanders might appear not as objects of study, offering or concealing knowledge from writers and researchers, but as speaking subjects authoring their own stories. In imagining futures for scholarship on and with them, I think often of linguist Elwyn Flint's improvised field practice, described in chapter 5. Encountering the detritus and reverberations of previous investigations, he collaborated with his partners to develop new recording techniques—though he also found that these were most efficacious when he left the room entirely. It is difficult for settler scholars like myself to imagine leaving the room, to imagine that our interventions are not needed, or to imagine the destruction of boundaries between studier and studied.[40] But with the cliff face gone and with the myth of the natural laboratory consigned to the dustbin, one wonders what other stories, visions, and voices, heretofore unseen or unimaginable in scholarship, might come into view. And then, perhaps, we may even begin to wonder whether scrutinizing visions or experimentalizing fantasies were needed in the first place.[41]

Notes

Introduction

1. Harry L. Shapiro, *The Heritage of the Bounty: The Story of Pitcairn through Six Generations* (Simon and Schuster, 1936), 139.

2. Matthew Spriggs, "Are Islands Islands? Some Thoughts on the History of Chalk and Cheese," in *Islands of Inquiry: Colonisation, Seafaring and the Archaeology of Maritime Landscapes*, ed. Geoffrey Clark et al., Terra Australis 29 (Australian National University Press, 2008), 211–226, https://www.jstor.org/stable/j.ctt24h8gp.16; Patrick V. Kirch, "Microcosmic Histories: Island Perspectives on 'Global' Change," *American Anthropologist* 99, no. 1 (1997): 30–42, https://doi.org/10.1525/aa.1997.99.1.30.

3. Letter from Lord Chelmsford, July 20, 1910, 1, the National Archives of the United Kingdom, Kew (hereafter TNA), CO 537/463.

4. Letter from A. W. Moverley to the Colonial Office, August 23, 1953, TNA, CO 1023/194.

5. Hardwicke Knight, "Preface," *Preliminary Report*, 1, Papers of Hardwicke Knight, Hocken Library, University of Otago, MS-4118-107.

6. See especially Greg Dening, *Mr Bligh's Bad Language: Passion, Power and Theatre on the Bounty* (Cambridge University Press, 1994), 340.

7. Terence Lewin, *Mutiny on the Bounty, 1789–1989: An International Exhibition to Mark the 200th Anniversary, 28 April 1989–1 Oct. 1989* (published by Manorial Research PLC in association with the National Maritime Museum, 1989), 4.

8. S. Gordon Gurwit, "Out of the Coral Sea," *Argosy All-Story Weekly* 147, no. 2 (November 18, 1922): 186.

9. "The Mutineers of the *Bounty*," *Literary World* 1, no. 2 (April 1, 1871): 1.

10. Charles Nordhoff and James Norman Hall, *The Mutiny on the* Bounty (Little, Brown, 1932); *Men against the Sea* (Little, Brown, 1934); and *Pitcairn's Island* (Little, Brown, 1934).

11. Sylvie Largeaud-Ortega, "Nordhoff and Hall's *Mutiny on the Bounty:* A Piece of Colonial Historical Fiction," in *The Bounty from the Beach: Cross-Cultural and Cross-Disciplinary Essays*, ed. Sylvie Largeaud-Ortega (Australian National University Press, 2018), 225–262, https://doi.org/10.22459/BB.10.2018.

12. Douglas Thorsen, "Only on Pitcairn," unpublished manuscript, 1982, Alexander Turnbull Library, National Library of New Zealand, Wellington (hereafter NLNZ), MS-Papers-3926.

13. *The Mutiny on the Bounty,* directed by Raymond Longford (Crick and Jones, 1916), likely lost; *In the Wake of the Bounty,* directed by Charles Chauvel (Expeditionary Films, 1933); *Mutiny on the Bounty,* directed by Frank Lloyd (Metro-Goldwyn-Mayer [MGM], 1935); *Mutiny on the Bounty,* directed by Lewis Milestone (MGM, 1962); *The Bounty,* directed by Roger Donaldson (Orion Pictures, 1984).

14. Dening, *Mr Bligh's Bad Language*; Caroline Alexander, *The Bounty: The True Story of the Mutiny on the* Bounty (Viking, 2003).

15. Henry Evans Maude, "In Search of a Home: From the Mutiny to Pitcairn Island (1789–1790)," *Journal of the Polynesian Society* 67, no. 2 (June 1958): 104–131.

16. I use "abducted" deliberately; the mutineers invited guests aboard and departed without warning. See Dening, *Mr Bligh's Bad Language,* 316.

17. Philip Carteret, "Captain Carteret's Voyage," in *An Account of the Voyages Undertaken by the Order of His Present Majesty for Making Discoveries in the Southern Hemisphere and Successively Performed by Commodore Byron, Captain Carteret, Captain Wallis, and Captain Cook in the Dolphin, the Swallow, and the Endeavour,* ed. John Hawkesworth (Strahan and Cadell, 1773), 1:561.

18. In addition to Dening, see Trevor Lummis, *Pitcairn Island: Life and Death in Eden* (Ashgate, 1997).

19. John Adams, "Statement to the Officers of the Blossom," 1825, Caird Library, National Maritime Museum (hereafter NMM), Greenwich, UK, NMM/PGR/14; Philip Pipon, "Narrative of the State Mutineers of H.M. Ship *Bounty* Settled on Pitcairn's Island in the South Seas," September 1814, State Library of New South Wales (hereafter SLNSW), Papers of Joseph Banks, 71.05; H. B. Willis, "Pitcairn's Island in the Southern Ocean," 1814, NLNZ, MS-2259; John Shillibeer, *A Narrative of the Briton's Voyage, to Pitcairn's Island* (Law and Whittaker, 1817).

20. Teehuteatuaonoa (Jenny), "Account of the Mutineers of the Ship *Bounty* and Their Descendants at Pitcairn's Island," *Sydney Gazette and New South Wales Advertiser* (July 17, 1819): 817, http://nla.gov.au/nla.news-article2178824; Teehuteatuaonoa (Jenny), "Account of Jenny, an Otahetian Woman," *United Services Journal* 2, no. 2 (November 1829): 589–593.

21. Epeli Hau'ofa, "Our Sea of Islands," *Contemporary Pacific* 6, no. 1 (1994): 148–161.

22. Alan S. C. Ross and A. W. Moverley, "Glossary," in *The Pitcairnese Language* (Andre Deutsch, 1964), 259.

23. Linda Colley, *The Gun, the Ship and the Pen: Warfare, Constitutions and the Making of the Modern World* (Liveright, 2021).

24. Lorenz Gonschor, "Norfolk Island," *Contemporary Pacific* 29, no. 1 (2017): 154–165, https://doi.org/10.1353/cp.2017.0011.

25. Dawn Oliver, *Justice, Legality, and the Rule of Law: Lessons from the Pitcairn Prosecutions* (Oxford University Press, 2009).

26. I have employed scare quotes to cordon off some words and ideas employed by outsiders when describing Pitcairn and Norfolk Islanders; I want to ensure a clear separation between notions held by historical actors and my own analysis of them.

27. Elizabeth Hennessy, "The Politics of a Natural Laboratory: Claiming Territory and Governing Life in the Galápagos Islands," *Social Studies of Science* 48, no. 4 (August 2018): 483–506, https://doi.org/10.1177/0306312718788179; Warwick Anderson, "Racial Hybridity, Physical Anthropology, and Human Biology in the Colonial Laboratories of the United States," *Current Anthropology* 53, no. S5 (April 2012): S95–107, https://doi.org/10.1086/662330; Michael Guggenheim, "Laboratizing and De-Laboratizing the World: Changing Sociological Concepts for Places of Knowledge Production," *History of the Human Sciences* 25, no. 1 (February 1, 2012): 99–118, https://doi.org/10.1177/0952695111422978; Helen Tilley, *Africa as Living Laboratory: Empire, Development, and the Problem of Scientific Knowledge* (University of

Chicago Press, 2011); Tony Bennett, "Museum, Field, Colony," *Journal of Cultural Economy* 2, no. 1–2 (July 1, 2009): 99–116, https://doi.org/10.1080/17530350903064097; Robert E. Kohler, *Landscapes and Labscapes: Exploring the Lab-Field Border in Biology* (University of Chicago Press, 2002); Henrika Kuklick and Robert E. Kohler, *Science in the Field*, Osiris 11 (University of Chicago Press, 1996); Henrika Kuklick, "Islands in the Pacific: Darwinian Biogeography and British Anthropology," *American Ethnologist* 23, no. 3 (1996): 611–638, https://doi.org/10.1525/ae.1996.23.3.02a00090; Richard H. Grove, *Green Imperialism: Colonial Expansion, Tropical Island Edens and the Origins of Environmentalism, 1600–1860* (Cambridge University Press, 1996).

28. Tamar I. Rozett, "The Case of Claud Cardew's Violin: Race, Anxiety, and the British Empire Mail," *Journal of British Studies* 61, no. 3 (July 2022): 599–620, https://doi.org/10.1017/jbr.2022.53; Joanna Lewis, *Empire of Sentiment: The Death of Livingstone and the Myth of Victorian Imperialism* (Cambridge University Press, 2018); Rachel King, "Living on Edge: New Perspectives on Anxiety, Refuge and Colonialism in Southern Africa," *Cambridge Archaeological Journal* 27, no. 3 (August 2017): 533–551, https://doi.org/10.1017/S0959774317000312; Michael Brown, "Cold Steel, Weak Flesh: Mechanism, Masculinity and the Anxieties of Late Victorian Empire," *Cultural and Social History* 14, no. 2 (March 15, 2017): 155–181, https://doi.org/10.1080/14780038.2016.1269538; Kenton Storey, *Settler Anxiety at the Outposts of Empire: Colonial Relations, Humanitarian Discourses, and the Imperial Press* (University of British Columbia Press, 2016); Kim A. Wagner, "'Treading upon Fires': The 'Mutiny'-Motif and Colonial Anxieties in British India," *Past and Present* 218, no. 1 (February 1, 2013): 159–197, https://doi.org/10.1093/pastj/gts026; Ann Laura Stoler, *Along the Archival Grain: Epistemic Anxieties and Colonial Common Sense* (Princeton University Press, 2010); Eric Allina, "'Fallacious Mirrors': Colonial Anxiety and Images of African Labor in Mozambique, ca. 1929," *History in Africa* 24 (1997): 9–52, https://doi.org/10.2307/3172017.

29. Stoler, *Along the Archival Grain*, 1.

30. While many historians have analyzed anxiety, derangement, in the sense I deploy it, is understudied. That said, an important point of departure for thinking about the generative epistemic consequences of frustration, madness, and colonial encounter is Johannes Fabian, *Out of Our Minds: Reason and Madness in the Exploration of Central Africa* (University of California Press, 2000).

31. Kuklick and Kohler, *Science in the Field;* Kohler, *Landscapes and Labscapes.*

32. For dystopic readings of Pitcairn, see Brandon Presser, *The Far Land: 200 Years of Murder, Mania, and Mutiny in the South Pacific* (PublicAffairs, 2022); Kathy Marks, *Lost Paradise: From Mutiny on the Bounty to a Modern-Day Legacy of Sexual Mayhem, the Dark Secrets of Pitcairn Island Revealed* (Simon and Schuster, 2009); Dea Birkett, *Serpent in Paradise* (Anchor Books, 1997). For Norfolk, see Robert Macklin, *Dark Paradise: Norfolk Island; Isolation, Savagery, Mystery and Murder* (Hachette Australia, 2013); Tim Latham, *Norfolk: Island of Secrets; The Mystery of Janelle Patton's Death* (Allen and Unwin, 2005).

33. Michel Foucault, *The Order of Things* (Vintage Books, 1971); see also discussions in Mariangela Palladino and John Miller, eds., *The Globalization of Space* (Pickering and Chato, 2015).

34. *Oxford English Dictionary*, "derange, v.," accessed October 10, 2022, http://www.oed.com/view/Entry/50545.

35. Anna Lowenhaupt Tsing, "From the Margins," *Cultural Anthropology* 9, no. 3 (August 1994): 279–297; Erik Harms et al., "Remote and Edgy: New Takes on Old Anthropological Themes," *HAU: Journal of Ethnographic Theory* 4, no. 1 (June 2014): 361–381, https://doi.org/10.14318/hau4.1.020; Andersson Saxer and Martin Ruben, "The Return of Remoteness: Insecurity, Isolation and Connectivity in the New World Disorder," *Social Anthropology* 27, no. 2 (2019): 140–155, https://doi.org/10.1111/1469-8676.12652; Shafqat Hussain,

Remoteness and Modernity: Transformation and Continuity in Northern Pakistan (Yale University Press, 2015). For remoteness and Pitcairn, see Maria Amoamo, "Remoteness and Myth Making: Tourism Development on Pitcairn Island," *Tourism Planning and Development* 8, no. 1 (February 2011): 1–19, https://doi.org/10.1080/21568316.2011.554035.

36. Adrian Young et al., "When Margins Are Centres: De-Ranging Pitcairn Island's Place in Pacific Scholarship," *Journal of the Polynesian Society* 130, no. 3 (September 30, 2021): 197–226; Joshua Nash, "Inside(r)-Outside(r): Toward a Linguistics and Sociology of Space on Pitcairn Island," *Australian Folklore* 31 (December 2, 2016), http://journals.kvasirpublishing.com/af/article/view/372.

37. Sujit Sivasundaram, *Islanded: Britain, Sri Lanka, and the Bounds of an Indian Ocean Colony* (University of Chicago Press, 2013).

38. Herbert Ford, *Pitcairn Island as a Port of Call: A Record, 1790–2010,* 2nd ed. (McFarland, 2012).

39. Paul Rabinow, *Reflections on Fieldwork in Morocco* (University of California Press, 1977); George W. Stocking, *Observers Observed: Essays on Ethnographic Fieldwork* (University of Wisconsin Press, 1984); James Clifford and George E. Marcus, *Writing Culture: The Poetics and Politics of Ethnography* (University of California Press, 1986).

40. Bruno Latour, *Science in Action: How to Follow Scientists and Engineers through Society,* (Harvard University Press, 1988). For the ongoing conversation on extraction and circulation of global knowledge, see Paige Madison, "All Things Bleak and Bare beneath a Brazen Sky: Practice and Place in the Analysis of Australopithecus," *History and Philosophy of the Life Sciences* 41, no. 2 (April 23, 2019): 19, https://doi.org/10.1007/s40656-019-0258-x; Warwick Anderson, "Marking Global Health History: The Postcolonial Worldliness of Biomedicine," *Social History of Medicine* 27, no. 2 (2014): 372–384; Neil Safier, "Global Knowledge on the Move: Itineraries, Amerindian Narratives, and Deep Histories of Science," *Isis* 101, no. 1 (March 2010): 133–145, https://doi.org/10.1086/652693; Kapil Raj, *Relocating Modern Science: Circulation and the Construction of Knowledge in South Asia and Europe, 1650–1900* (Palgrave Macmillan, 2007); James A. Secord, "Knowledge in Transit," *Isis* 95, no. 4 (December 1, 2004): 654–672, https://doi.org/10.1086/430657.

41. Eve Tuck and K. Wayne Yang, "R-Words: Refusing Research," in *Humanizing Research: Decolonizing Qualitative Inquiry with Youth and Communities*, ed. Django Paris and Maisha T. Winn (Sage, 2013), 223–248; Rosanna Dent, "Whose Home Is the Field?," *Isis* 113, no. 1 (March 2022): 137–143, https://doi.org/10.1086/718150.

42. Margaret M. Bruchac, *Savage Kin: Indigenous Informants and American Anthropologists* (University of Arizona Press, 2018).

43. James C. Scott, *Domination and the Arts of Resistance: Hidden Transcripts* (Yale University Press, 2008); Robert Proctor and Londa L. Schiebinger, eds., *Agnotology: The Making and Unmaking of Ignorance* (Stanford University Press, 2008).

44. Adam Johnson, "Secretsharers: Intersecting Systems of Knowledge and the Politics of Documentation in Southwesternist Anthropology, 1880–1930" (PhD diss., University of Michigan, 2018), http://deepblue.lib.umich.edu/handle/2027.42/146056; Tony Crook, *Anthropological Knowledge, Secrecy and Bolivip, Papua New Guinea: Exchanging Skin* (Oxford University Press, 2007).

45. Moshe Rapaport, *The Pacific Islands: Environment and Society,* rev. ed. (University of Hawai'i Press, 2013), x.

46. B. P. McEvoy et al., "European and Polynesian Admixture in the Norfolk Island Population," *Heredity* 105, no. 2 (2010): 229–234, https://doi.org/10.1038/hdy.2009.175; Stuart Macgregor et al., "Legacy of Mutiny on the *Bounty:* Founder Effect and Admixture on Norfolk Island," *European Journal of Human Genetics* 18 (2010): 67–72.

47. Hardwicke Knight, *Pitcairn Survey Report* (1964), 9–19, Hardwicke Knight Papers (hereafter Knight Papers), Misc-MS-339.

48. Untitled image and photographs, in folder "Petroglyphs Pitcairn," Knight Papers, MS-4118–157.

49. "Interview with Helen Leach," by Meg Davidson, March 4, 2011, cited in her unpublished biography of Hardwicke Knight.

50. Interview with Lin Phelan and Linden Cowell by the author, April 19, 2014, Dunedin, New Zealand.

51. Rhiannon Adam, *Big Fence* (Blow Up Press, 2022).

Chapter 1: Eden Refounded

1. Montague Corri, *A New Romantick Operatick Ballet Spectacle, Founded on the Recent Discovery of a Numerous Colony, Formed by, and Descended from the Mutineers of the Bounty Frigate Called Pitcairn's Island* (London, 1816), http://nla.gov.au/nla.obj-33744670 (hereafter *Pitcairn's Island*).

2. Corri, *Pitcairn's Island,* 7.

3. Corri, *Pitcairn's Island,* 10.

4. Corri, *Pitcairn's Island,* 14.

5. I will use "British" and "English" interchangeably, reflecting the mixed use of both categories by my actors. The project of building British nationhood was then, as now, an unfinished one. See Krishnan Kumar, *The Making of English National Identity* (Cambridge University Press, 2003); Linda Colley, *Britons: Forging the Nation 1707–1837* (Yale University Press, 1992).

6. Bronwen Douglas, *Science, Voyages, and Encounters in Oceania, 1511–1850* (Palgrave Macmillan, 2014); Vanessa Smith, *Intimate Strangers: Friendship, Exchange and Pacific Encounters* (Cambridge University Press, 2010); Harry Liebersohn, *The Travelers' World: Europe to the Pacific* (Harvard University Press, 2006); Rod Edmond, *Representing the South Pacific: Colonial Discourse from Cook to Gauguin* (Cambridge University Press, 1997); Neil Rennie, *Far-Fetched Facts: The Literature of Travel and the Idea of the South Seas* (Clarendon, 1995); Greg Dening, *Islands and Beaches: Discourse on a Silent Land: Marquesas, 1774–1880* (Dorsey Press, 1980).

7. See especially Jonathan Lamb, *Preserving the Self in the South Seas, 1680–1840* (University of Chicago Press, 2001).

8. See Kathleen Wilson, *The Island Race: Englishness, Empire and Gender in the Eighteenth Century* (Routledge, 2003); Robert Young, *Colonial Desire: Hybridity in Theory, Culture, and Race* (Routledge, 1995); Mary Louise Pratt, *Imperial Eyes: Travel Writing and Transculturation* (Routledge, 1992); Homi Bhabha, *The Location of Culture* (Routledge, 1994).

9. Simon Gikandi, *Maps of Englishness: Writing Identity in the Culture of Colonialism* (Columbia University Press, 1996). Jonathan Lamb, in *Preserving the Self in the South Seas,* analyzes the use of litotes in Pacific travel narratives: e.g., "the natives are not unhandsome."

10. Young et al., "When Margins Are Centres," 197–226.

11. "1808," *Pitcairn Island Register,* NMM, REC 61. The *Register*'s early entries were written years after the arrival of the *Topaz.*

12. Amasa Delano, *Narrative of Voyages and Travels in the Northern and Southern Hemispheres: Comprising Three Voyages round the World; Together with a Voyage of Survey and Discovery, in the Pacific Ocean and Oriental Islands* (E. G. House, 1817). Delano was a captain, too, and lives on in literature as a central character in Herman Melville's *Benito Cereno* (1855).

13. Delano, *Narrative of Voyages and Travels,* 139–140. Dialogue has been reformatted from the original.

14. Delano, *Narrative of Voyages and Travels,* 139.

15. Reprinted widely, including two articles in the *Quarterly Review:* "Voyage de Dentrecasteaux," *Quarterly Review* 3, no. 5 (February 1810): 21–43; "Journal of a Cruise Made to the Pacific Ocean, by Captain David Porter," *Quarterly Review* 13, no. 26 (July 1815): 374–383. Both articles are anonymous but presumably by Sir John Barrow, who was second secretary to the Admiralty.

16. "Voyage de Dentrecasteaux," 24.

17. John Barrow, *Interesting Narratives and Discoveries Including the Life of James Bruce, Esquire, Travels in South Africa, by J. Barrow, Esq., Discovery of an English Colony on Pitcairn's Island in the South Sea* (C. Hulbert, 1817); Delano, *Narrative of Voyages and Travels;* Shillibeer, *A Narrative of the Briton's Voyage;* Charles Lenox Sargent and John Adams, *The Life of Alexander Smith, Captain of the Island of Pitcairn, One of the Mutineers on Board His Majesty's Ship Bounty, Commanded by Lieut. Wm. Bligh* (S. T. Goss, 1819); "Descendants of the *Bounty*'s Crew, on Pitcairn's Island," *Calcutta Journal* 4, no. 133 (July 13, 1819).

18. Mary Russell Mitford, *The Life of Mary Russell Mitford: Told by Herself in Letters to Her Friends* (Harper and Brothers, 1870); Alison Booth, "Revisiting the Homes and Haunts of Mary Russell Mitford," *Nineteenth-Century Contexts* 30, no. 1 (2008): 39–65.

19. Mary Russell Mitford, *Christina, the Maid of the South Seas; a Poem* (A. J. Valpy, 1811).

20. Mary Russell Mitford, *Our Village: Sketches of Rural Character and Scenery* (E. Bliss, 1828), first published as a serialization in *Lady's Magazine.* Literary scholars have produced some readings of *Christina:* see Catherine Addison, "Gender and Race in Byron's *The Island* and Mitford's *Christina,*" *Antipodes* 18, no. 1 (June 2004): 72–76; Elisa E. Beshero-Bondar performed a close reading to suggest "musical performance poses a Romantic tension between verbal authority and nonverbal resistance" in "Romancing the Pacific Isles before Byron: Music, Sex, and Death in Mitford's *Christina,*" *ELH* 76, no. 2 (2009): 276–308. Beshero-Bondar is the principal editor of Digital Mitford: The Mary Russell Mitford Archive, a growing digital history project (http://www.pitt.edu/~ebb8/DigitalMitford/index.html). Byron's poem is *The Island, or Christian and His Comrades* (John Hunt, 1823).

21. "The sudden shock / scatter'd his brains upon the rock." M. R. Mitford, *Christina,* canto III, verse xxi.

22. M. R. Mitford, *Christina,* p. vi.

23. M. R. Mitford, *Christina,* pp. vi–vii.

24. From her citations, it is possible to partially reconstruct Mitford's Pacific library, which includes William Bligh, *A Voyage to the South Sea Undertaken by Command of His Majesty* (1792); James Cook, *The Voyages of Captain James Cook round the World* (1809); George Vason, *An Authentic Narrative of Four Years' Residence at Tongataboo* (1810); James Burney, *A Chronological History of the Discoveries in the South Sea or Pacific Ocean* (1803); Alexander Dalrymple, *An Historical Collection of the Several Voyages and Discoveries in the South Pacific Ocean* (1770); John Hawkesworth, *An Account of the Voyages Undertaken by the Order of His Present Majesty for Making Discoveries in the Southern Hemisphere* (1773); William Dampier, *A New Voyage round the World* (1703); and John Fegan, *Voyage around the World by Lewis de Bougainville 1766–9.*

25. "Voyage de Dentrecasteaux."

26. M. R. Mitford, *Christina,* canto I, pp. 206–208, n. 9, citing Cook, *The Voyages of Captain James Cook,* 221–224. Yes, these are footnotes to footnotes.

27. M. R. Mitford, *Christina,* canto II, pp. 230–231, n. 7, citing Cook, in Hawkesworth, *An Account of the Voyages,* 187–188.

28. M. R. Mitford, *Christina,* canto I, pp. 208–210, n. 10; canto II, p. 300, n. 22, citing Hawkesworth, *An Account of the Voyages,* 2:204; canto II, pp. 225–228, n. 5, citing Bligh, *A Voyage to the South Sea,* 65, and Alvera Mendana de Neyra, quoted in Dalrymple, *An Historical Collection,* 81.

29. M. R. Mitford, *Christina,* canto III, p. 313, n. 7, citing Bligh, *A Voyage to the South Sea,* 64.

30. M. R. Mitford, *Christina,* canto I, pp. 200–202, n. 4, citing Banks' and Solander's description in Cook, *The Voyages of Captain James Cook.*

31. M. R. Mitford, *Christina,* canto III, pp. 313–318, n. 8, citing Folger's letter in the *Quarterly Review.*

32. Though Hubert may have been modeled loosely on Thursday October Christian.

33. M. R. Mitford, *Christina,* canto III, p. 318, n. 8.

34. M. R. Mitford, *Christina,* pp. v–vi.

35. Denis Diderot, "Supplément au voyage de Bougainville" [Addendum to the journey of Bougainville], in *Diderot: Political Writings,* ed. John Hope Mason and Robert Wokler, Cambridge Texts in the History of Political Thought (Cambridge University Press, 1992), 31–76.

36. M. R. Mitford, *Christina,* canto I, verse ix.

37. M. R. Mitford, *Christina,* canto I, verse xiii.

38. M. R. Mitford, *Christina,* canto III, verse xvi.

39. For more on race and colonial notions of transplantation, see Suman Seth, *Difference and Disease: Medicine, Race, and the Eighteenth-Century British Empire* (Cambridge University Press, 2018).

40. Shillibeer, *A Narrative of the Briton's Voyage;* Philip Pipon, *The Descendants of the Bounty's Crew* (Henry Colburn, 1834); captain of the *Briton* Sir Thomas Staines's account was first published alongside Folger's letters, presumably by Barrow, in "Journal of a Cruise Made to the Pacific Ocean by Captain David Porter."

41. Pipon, "Narrative of the State Mutineers of H.M. Ship *Bounty,*" 9–10.

42. Pipon, "Narrative of the State Mutineers of H.M. Ship *Bounty,*" 6.

43. H. B. Willis, "Pitcairn's Island in the Southern Ocean," 1814, NLNZ, qMS-2259.

44. John Lawrie, "Pitcairn Island, 1822 Aboard *Hercules,*" 95–95, National Library of Australia (hereafter NLA), MS 9122.

45. Lawrie, "Pitcairn Island, 1822," 105.

46. Lawrie, "Pitcairn Island, 1822," 113 (emphasis added).

47. Frederick William Beechey, *Narrative of a Voyage to the Pacific and Beering's Strait, to Co-operate with the Polar Expeditions, Performed in His Majesty's Ship "Blossom," under the Command of Captain F. W. Beechey in the Years 1825, 26, 27, 28* (Henry Colburn and Richard Bentley, 1831); George Peard, *The Journal of Lieutenant George Peard of H.M.S. "Blossom," 1825–1828,* ed. Barry M. Gough (Cambridge University Press, 1973); Edward Belcher, "Private Journal, remarks etc. HM Ship Blossom on discovery during the years 1825, 6, 7, Capt F W Beechey Comdr," unpublished manuscript, NLNZ, MS-0158; James Wolfe, *Journal of Voyage in HMS Blossom,* NLNZ, Micro-MS-Coll-08-0538.

48. Peard, *Journal,* 75–76.

49. Peard, *Journal,* 76. There were rumors in Britain that Christian survived and had secreted himself back to his native country; pardoned mutineer Peter Heywood claimed to have seen him disappear around a street corner. Heywood's story is reproduced in his stepdaughter's history: Diana Jolliffe Belcher, *The Mutineers of the Bounty and Their Descendants in Pitcairn and Norfolk Islands* (Harper and Brothers, 1871), 175.

50. Beechey, *Narrative of a Voyage,* 66.

51. Beechey, *Narrative of a Voyage,* 67.

52. Beechey, *Narrative of a Voyage,* 67.

53. Adams's Statement to the Officers of the *Blossom,* 1825, NMM/PGR/14.

54. Beechey, *Narrative of a Voyage,* 94.

55. Shillibeer, *A Narrative of the Briton's Voyage,* 88–89; Beechey, *Narrative of a Voyage,* 102.

56. Beechey, *Narrative of a Voyage,* 121–122.

57. Beechey, *Narrative of a Voyage,* 105.

58. Beechey, *Narrative of a Voyage,* 100, 104.

59. Beechey, *Narrative of a Voyage*, 111–112. Dancing remained essentially forbidden on Pitcairn for the next two centuries except on select occasions.

60. Peard, *Journal*, 93.

61. Beechey, *Narrative of a Voyage*, 111–112.

62. Beechey, *Narrative of a Voyage*, 127–128.

63. Shillibeer, *A Narrative of the Briton's Voyage*, 82–83.

64. Beechey, *Narrative of a Voyage*, 124.

65. Beechey, *Narrative of a Voyage*, 122.

66. Christopher Lloyd, *Mr. Barrow of the Admiralty: A Life of Sir John Barrow, 1764–1848* (Collins, 1970); John Barrow, *An Auto-biographical Memoir of Sir John Barrow, Bart., Late of the Admiralty: Including Reflections, Observations, and Reminiscences at Home and Abroad, from Early Life to Advanced Age* (John Murray, 1847).

67. He contributed hundreds of articles to the journal, many of them anonymous articles that have now been attributed: see Jonathan Cutmore, "The Early 'Quarterly Review': New Attributions of Authorship," *Victorian Periodicals Review* 28, no. 4 (December 1, 1995): 305–329.

68. John Barrow, *The Eventful History of the Mutiny and Piratical Seizure of H. M. S. Bounty, Its Causes and Consequences* (J. Murray, 1831).

69. Barrow, *The Eventful History*, 183; cf. John Barrow, "Spirit of Discovery," *Mirror* 23, no. 647 (February 8, 1834): 83–86.

70. Abraham Thompson Lowe, *Second Class Book: Principally Consisting of Historical, Geographical, and Biographical Lessons, Adapted to the Capacities of Youth, and Designed for Their Improvement: Intended for the Use of Schools* (Dorr and Howland, 1829), 102.

71. Sylvester Graham, *A Lecture to Young Men on Chastity: Intended Also for the Serious Consideration of Parents and Guardians* (G. W. Light, 1838), 152.

72. *The Ship and the Island* (Groombridge and Sons, 1849).

73. The story of John Adams's conversion, for example, is essentially a line-by-line paraphrase of Barrow's prose. Compare pp. 33–34 in *The Ship and the Island* to pp. 306–307 in Barrow's *Eventful History*.

74. N. W. Fiske, *Aleck: The Last of the Mutineers, or the History of Pitcairn's Island* (E. C. Biddle, 1845).

75. Fiske, *Aleck*, iii.

76. Fiske, *Aleck*, 155.

77. Fiske, *Aleck*, 134–135.

78. Thomas Boyles Murray, *Pitcairn, the Island, the People, and the Pastor* (Society for Promoting Christian Knowledge, 1857).

79. Murray, *Pitcairn*, xiv.

80. John Barrow, *A Description of Pitcairn's Island and Its Inhabitants: With an Authentic Account of the Mutiny of the Ship Bounty and of the Subsequent Fortunes of the Mutineers* (J. and J. Harper, 1832), 278–284.

81. Murray, *Pitcairn*, 112.

82. *The Converted Mutineer and His Bible Class: Or, John Adams and the Children of the Mutineers* (Massachusetts Sabbath School Society, 1855); Thomas Boyles Murray, *The Home of the Mutineers* (American Sunday-School Union, 1854); *The Transformed Island: a Story of the South Seas* (Presbyterian Board of Publication, 1854).

83. *Collector* 6, no. 15 (June 1, 1895): 249; also published as "Relics of the Past in London," *Nation* 60, no. 1558 (May 9, 1895): 357–358. The lock of hair and its frame are now part of the collections of the National Maritime Museum in Greenwich, "Pigtail," NMM/REL0003.

84. Letter from Fortescue Moresby to unknown recipient, August 1852, reprinted, NLA, Rex Nan Kivell Collection, NK2295, http://nla.gov.au/nla.gen-nk2295. Also cited in Murray, *Pitcairn*, 211.

85. See Nicholas Thomas, *Out of Time: History and Evolution in Anthropological Discourse* (University of Michigan Press, 1989); Johannes Fabian, *Time and the Other: How Anthropology Makes Its Object* (Columbia University Press, 2002).

86. "Memorabilia of the Exhibition Season," *Fraser's Magazine* 40, no. 260 (August 1851): 1–2.

87. See Jeffrey A. Auerbach, *The Great Exhibition of 1851: A Nation on Display* (Yale University Press, 1999).

88. "Memorabilia of the Exhibition Season," 2.

89. The exhibition of people from other cultures in human zoos has a very long history. See Sadiah Qureshi, *Peoples on Parade: Exhibitions, Empire, and Anthropology in Nineteenth-Century Britain* (University of Chicago Press, 2011).

90. "A Pitcairn in London," *Launceston Examiner*, July 7, 1881, 3.

91. The story was retold in Rosalind Amelia Young, *Mutiny of the Bounty and Story of Pitcairn Island, 1790–1894* (Pacific Press, 1894), 219.

92. "Pitcairn Island," *Penny Illustrated Paper and Illustrated Times*, no. 1034 (May 7, 1881).

93. Letter from John Barrow to R. W. Hay, November 29, 1826, TNA, FO 58/14.

94. *Canton Register* (January 16, 1832), cited in H. E. Maude, "Tahitian Interlude: The Migration of the Pitcairn Islanders to the Motherland in 1831," *Journal of the Polynesian Society* 68, no. 2 (1959): 127.

95. John Barrow et al., "Recent Accounts of the Pitcairn Islanders," *Journal of the Royal Geographical Society of London* 3 (January 1833): 156–162, https://doi.org/10.2307/1797603.

96. Cf. Maude, *Tahitian Interlude*, 124.

97. Barrow et al., "Recent Accounts of the Pitcairn Islanders," 163–164.

98. Maude, *Tahitian Interlude*, 134.

99. Pauline Reynolds, "Tapa Cloths and Beaters: Tradition, Innovation and the Agency of the Bounty Women in Shaping a New Culture on Pitcairn Island from 1790 to 1850," *Textile History* 47, no. 2 (July 2, 2016), 191–193, https://doi.org/10.1080/00404969.2016.1211435.

100. Letter from A. A. Sandilands of HMS *Comet*, April 7, 1831, TNA, FO 58/6.

101. Cf. Pipon, *The Descendants of the Bounty's Crew*; Belcher, *The Mutineers of the Bounty*.

102. Barrow et al., "Recent Accounts of the Pitcairn Islanders," 166.

103. William Gunn, *Medical Journal of the HMS Curacoa* (1841), TNA, ADM/101/95/4A.

104. Samuel Greatheed—Authentic History of the Mutineers of the *Bounty* (1820), NLNZ, Micro-MS-Coll-08-0099.

105. John Buffett, "A Narrative, of 20 Years' Residence on Pitcairn's Island," *Friend* 4, no. 1 (1846): 2–3.

106. Buffett, "A Narrative, of 20 Years' Residence on Pitcairn's Island."

107. Buffett, "A Narrative, of 20 Years' Residence on Pitcairn's Island."

108. Letter from Nobbs to Fairfax Moresby, reprinted in Belcher, *The Mutineers of the Bounty*, 233.

109. Barrow, *The Eventful History of the Mutiny*, 330–331.

110. Much of this account is necessarily drawn from texts penned by Hill's rivals, especially Buffett's account.

111. Buffett, "A Narrative, of 20 Years' Residence on Pitcairn's Island."

112. The story is repeated in numerous sources, including Belcher, *The Mutineers of the Bounty*, 178.

113. E.g., Michelle Elleray, "Crossing the Beach: A Victorian Tale Adrift in the Pacific," *Victorian Studies* 47, no. 2 (Winter 2005): 164–173; Rod Edmond, "Mutineers and Beachcombers," in *Representing the South Pacific*, 63–97.

114. Tillman W. Nechtman, *The Pretender of Pitcairn Island: Joshua W. Hill—The Man Who Would Be King among the Bounty Mutineers* (Cambridge University Press, 2018).

Chapter 2: The "Experiment"

1. Despatch from John Pakington to William Denison, December 15, 1852, "Correspondence on the Subject of Removal of Inhabitants of Pitcairn's Island to Norfolk Island," *British Parliamentary Papers* (HMSO, 1857).

2. John Nihill, "Journal of Her Majesty's Sloop Cockatrice" (1852), 24, TNA, ADM/101/94/3B.

3. See Nancy Stepan, *The Idea of Race in Science: Great Britain, 1800–1960* (Macmillan, 1982); Alison Bashford, *Imperial Hygiene: A Critical History of Colonialism, Nationalism and Public Health* (Palgrave Macmillan, 2003); Warwick Anderson, *The Cultivation of Whiteness: Science, Health, and Racial Destiny in Australia* (Duke University Press, 2006); Bronwen Douglas and Chris Ballard, eds., *Foreign Bodies: Oceania and the Science of Race 1750–1940* (Australian National University Press, 2008); Damon Ieremia Salesa, *Racial Crossings: Race, Intermarriage, and the Victorian British Empire* (Oxford University Press, 2011); Chloe Campbell, *Race and Empire: Eugenics in Colonial Kenya* (Oxford University Press, 2012); Bronwen Douglas, *Science, Voyages, and Encounters in Oceania, 1511–1850* (Palgrave Macmillan, 2014).

4. For similar cases, see Kate Stevens, *Gender, Violence and Criminal Justice in the Colonial Pacific: 1880–1920* (Bloomsbury Publishing, 2022).

5. C. H. Currey, "Sir William Thomas Denison (1804–1871)," in *Australian Dictionary of Biography* (National Centre of Biography, Australian National University), published 2006, http://adb.anu.edu.au/biography/denison-sir-william-thomas-3394/text5145. Denison published an extensive two-volume autobiography, *Varieties of Vice-Regal Life* (Longmans, Green, 1870). One biographical account of Denison's rule in the antipodes exists: J. M. Bennett, *Reluctant Democrat: Sir William Denison in Australia 1847–1861* (Federation Press, 2011). The "extinction of the Tasmanians" was, like Pitcairn, a much-fetishized case in Victorian anthropology; see Patrick Brantlinger, *Dark Vanishings: Discourse on the Extinction of Primitive Races, 1800–1930* (Cornell University Press, 2003); Russell McGregor, "The Doomed Race: A Scientific Axiom of the Late Nineteenth Century," *Australian Journal of Politics and History* 39, no. 1 (April 1, 1993): 14–22.

6. See Maev O'Collins, *An Uneasy Relationship: Norfolk Island and the Commonwealth of Australia* (Australian National University Press, 2010), 1–18.

7. The *Oxford English Dictionary* has references for the meaning "an action or operation undertaken in order to discover something unknown, to test a hypothesis, or establish or illustrate some known truth," which extends at least to the fourteenth century. *Oxford English Dictionary*, "experiment, n," accessed May 4, 2016, http://www.oed.com/view/Entry/66530.

8. Despatch from William Denison to Lord John Russell, September 3, 1857, "Correspondence on the Subject of Removal of Inhabitants of Pitcairn's Island to Norfolk Island," *British Parliamentary Papers* (HMSO, 1857).

9. Letter from Nicholas Toup to the Pitcairn Islanders, July 5, 1854, "Correspondence on the Subject of Removal of Inhabitants of Pitcairn's Island to Norfolk Island."

10. Maude, "Tahitian Interlude."

11. Stephen Fremantle, "Visit to Pitcairn's Island," HMS *Juno*, September 23, 1855, TNA, CO 201/496.

12. Fremantle, "Visit to Pitcairn's island."

13. Instructions for Lieutenant Gregorie from William Denison, 1856, reproduced and enclosed in Despatch from William Denison to George Grey, February 27, 1856, "Correspondence on the Subject of Removal of Inhabitants of Pitcairn's Island to Norfolk Island."

14. Letter from G. W. Gregorie to William Denison, June 28, 1856, TNA, ADM 125/135.

15. Letter from Denison the Pitcairn Islanders, reproduced and enclosed in Despatch from William Denison to George Grey, February 27, 1856.

16. Letter from G. W. Gregorie to William Denison, June 28, 1856.

17. G. A. Gregorie, "April 22," *Journal of a Voyage from Sydney to Pitcairn and Norfolk Islands on Board the Morayshire,* February 23–June 8, 1856, TNA, RUSI/NM/138, hereafter *Journal.*

18. Gregorie, "April 23," *Journal.*

19. Gregorie, "April 24," *Journal.*

20. Gregorie, "April 27–28," *Journal.*

21. Gregorie, "May 3," *Journal.*

22. Gregorie, "May 22," *Journal.*

23. Letter from G. W. Gregorie to William Denison, June 28, 1856.

24. Gregorie, "June 8," *Journal.*

25. Letter from G. W. Gregorie to William Denison, June 28, 1856.

26. Letter from Stephen Fremantle to Ralph Osborne, Admiralty, "Visit to Norfolk Island," June 28, 1856, ADM 125/135.

27. Letter from G. W. Gregorie to William Denison, June 28, 1856.

28. John Buffett, "June 8," *Diary of John Buffett* (1856), Turnbull Library, Wellington, Micro-MS-Coll-08-0123.

29. Letter from Catherine Christian to Frederick Howard, October 8, 1857, Letters to Frederick Howard, KAVHA Museum, Norfolk Island.

30. Letter from Captain Fremantle, June 28, 1856, ADM 125/135.

31. "Letter to Labouchere from Denison, July 21, 1856," reproduced in Denison, *Varieties of Vice-Regal Life,* 395.

32. Duncan Bell, *Victorian Visions of Global Order: Empire and Relations in Nineteenth-Century Political Thought* (Cambridge University Press, 2007), 3.

33. John Barrow et al., "Recent Accounts of the Pitcairn Islanders"; "Royal Geographical Society," *Athenaeum,* no. 505 (July 1, 1837): 484.

34. Robert J. Mayhew, *Malthus: The Life and Legacies of an Untimely Prophet* (Harvard University Press, 2014); James P. Huzel, *The Popularization of Malthus in Early Nineteenth-Century England: Martineau, Cobbett and the Pauper Press* (Ashgate, 2006).

35. Thomas Robert Malthus, "Of the Checks to Population in the Islands of the South Sea," in *An Essay on the Principle of Population: Or, a View of Its Past and Present Effects on Human Happiness* (T. Bensley, 1807), 79–80.

36. Thomas Doubleday, *The True Law of Population Shewn to Be Connected with the Food of the People* (Simpkin, Marshall, 1842).

37. Review of *The True Law of Population,* by Thomas Doubleday, *British Quarterly Review* 2, no. 4 (August 1845): 442.

38. William Edward Hickson, *Malthus: An Essay on the Principle of Population in Refutation of the Theory of the Rev. T. R. Malthus* (Taylor, Walton, and Maberly, 1849), 72–73.

39. Henry George, *Progress and Poverty: An Inquiry into the Cause of Industrial Depressions and of Increase in Want with Increase in Wealth, the Remedy* (William Reeves, 1884), 80.

40. Lord Bramwell, *Nationalisation of Land: A Review of Mr. Henry George's "Progress and Poverty"* (Liberty and Property Defence League, 1882), 7 (emphasis in original).

41. Terence D. Keel, "Religion, Polygenism and the Early Science of Human Origins," *History of the Human Sciences* 26, no. 2 (April 1, 2013): 3–32; Robert Kenny, "From the Curse of Ham to the Curse of Nature: The Influence of Natural Selection on the Debate on Human Unity before the Publication of *The Descent of Man,*" *British Journal for the History of Science* 40, no. 3 (September 2007): 367–388; Adam Dewbury, "The American School and Scientific Racism in Early American Anthropology," *Histories of Anthropology Annual* 3, no. 1 (2007): 121–147; Kathy Alexis Psomiades, "Polygenist Ecosystems: Robert Knox's *The Races of Man* (1850)," *Victorian Review* 36, no. 2 (2010): 32–36; for surveys on scientific racism, see John P. Jackson and Nadine M. Weidman, *Race, Racism, and Science: Social Impact and Interaction*

(ABC-CLIO, 2004); Stepan, *The Idea of Race in Science;* and Stephen Jay Gould's popular *The Mismeasure of Man* (W. W. Norton, 1996), which is often cited but also often critiqued.

42. Josiah Nott, *Two Lectures on the Natural History of the Caucasian and Negro Races* (Dade and Thompson, 1844), 34; white, masculine insecurity obviously undergirded examples such as these; theorists never seemed especially eager to reverse the scenario and deposit white women with Black men on deserted islands.

43. "Unity of the Races," *Southern Quarterly Review* 7, no. 14 (April 14, 1845): 446.

44. The Ethnological Society was an offshoot of the Aborigines' Protection Society, which was itself an offshoot from the abolition movement. The ESL and ASL merged to form the Royal Anthropological Institute in 1871.

45. F. W. Farrah, "On Hybridity," *Journal of the Anthropological Society of London* 2 (1864): 222–224.

46. Farrah, "On Hybridity," 227.

47. Farrah, "On Hybridity," 227.

48. John Crawfurd, "On the Supposed Infecundity of Human Hybrids or Crosses," *Transactions of the Ethnological Society of London* 3 (January 1865): 356–362, https://doi.org/10.2307/3014171.

49. Thomas Kelly Cheyne, *The Relations between Civilized and Uncivilized Races: A Price Essay Read in the Theatre, Oxford, June 8, 1864* (T. and G. Shrimpton, 1864), 28.

50. Letter from Charles Darwin to Carline Denison, 1874, NLNZ, MS-Papers-0972.

51. Charles Darwin, *The Descent of Man and Sexual Selection in Relation to Sex* (John Murray, 1871), 154.

52. Alfred Russel Wallace, "The Mutineers of the Bounty, and Their Descendants in Pitcairn and Norfolk Islands, by Lady Belcher," *The Academy* (February 1, 1871): 108.

53. Wallace privileged the moral sphere, too, in demarcating the boundaries between human populations in his biogeographical work: Jeremy Vetter, "Wallace's Other Line: Human Biogeography and Field Practice in the Eastern Colonial Tropics," *Journal of the History of Biology* 39, no. 1 (March 2006): 89–123.

54. Wallace, "The Mutineers of the Bounty," 109.

55. Cf. Shino Konishi, "Idle Men: The Eighteenth-Century Roots of the Indigenous Indolence Myth," in *Passionate Histories: Myth, Memory, and Indigenous Australia,* ed. Frances Peters-Little et al., Aboriginal History Monographs 21 (Australian National University Press, 2010), 99–122; see also Barbara Arneil, *John Locke and America: The Defence of English Colonialism* (Clarendon Press, 1996); David Arnold, "'Illusory Riches': Representations of the Tropical World, 1840–1950," *Singapore Journal of Tropical Geography* 21, no. 1 (2000): 6–18; Felix Driver and Luciana Martins, *Tropical Visions in an Age of Empire* (University of Chicago Press, 2005).

56. Despatch from William Denison to Henry Labouchere, October 30, 1857, "Copy of Correspondence with the Government of New South Wales, in Reference to the Pitcairn Islanders Settled in Norfolk Island," *British Parliamentary Papers* (HMSO, 1863), 14; on Wakefield, see Edward R. Kittrell, "Wakefield's Scheme of Systematic Colonization and Classical Economics," *American Journal of Economics and Sociology* 32, no. 1 (1973): 87–111.

57. Belcher, *The Mutineers of the Bounty.* Original copies of these reports are at the National Archives; see especially CO 201/531, CO 201/577, CO 201/579, CO 537/462, and CO 537/463.

58. Despatch from John Young to the Duke of Newcastle, May 10, 1862, "Copy of Correspondence with the Government of New South Wales," 47–48.

59. Stoler, *Along the Archival Grain.*

60. *Papers Related to Her Majesty's Colonial Possessions, Reports for 1883, 1884, and 1885* (HMSO, 1886).

61. Letter from Viscount Hampden to Joseph Chamberlain, January 20, 1896, "Correspondence Relating to the Transfer of Norfolk Island to the Government of New South Wales," *British Parliamentary Papers* (HMSO, 1897), 4.

62. Enclosure by Viscount Hampden, December 2, 1895, "Correspondence Relating to the Transfer of Norfolk Island to the Government of New South Wales," 3.

63. Letter from Byron Adams and O. Masey Quintal to Viscount Hampden, February 24, 1896, "Correspondence Relating to the Transfer of Norfolk Island to the Government of New South Wales," 15.

64. William Houston, "Administration of Norfolk Island," June 13, 1903, CO 537/462.

65. "Extract from a resident of Norfolk Island (No 67) to H. L. Thompson, Secretary to the Deputy Administrator of Norfolk Island Affairs, written confidentially," May 12, 1903, CO 536/462.

66. Cf. Margaret Mishra, "'Your Woman Is a Very Bad Woman': Revisiting Female Deviance in Colonial Fiji," *Journal of International Women's Studies* 17, no. 4 (July 31, 2016): 67–78.

67. Anne Stoler, *Carnal Knowledge and Imperial Power: Race and the Intimate in Colonial Rule* (University of California Press, 2002); Bashford, *Imperial Hygiene;* Philippa Levine, *Prostitution, Race and Politics: Policing Venereal Disease in the British Empire* (Routledge, 2003).

68. "Visit of H. E. the Governor and the Deputy Administrator," *Appendix H: Notes of Interview with Constable Collins as to the Character of Occupants of Houses Owned by Government,* May 21, 1903, CO 536/462.

69. I have anonymized the names with dashes.

70. Houston, "Visit of H. E. the Governor and the Deputy Administrator," *Appendix D: Minutes of Proceedings,* undated, paragraphs 697–712, CO 536/462.

71. A passage itself cited by Lord Chelmsford in his 1910 report: see letter from Lord Chelmsford, July 20, 1910, 4, TNA, CO 537/463.

72. I have not been able to locate the list.

73. William Houston, "Administration of Norfolk Island," June 13, 1903, CO 537/462.

74. Houston, "Administration of Norfolk Island."

75. Letter from Lord Chelmsford, July 20, 1910, 1, CO 537/463 (hereafter Chelmsford Report).

76. Anderson, *The Cultivation of Whiteness,* 1–4, 14.

77. Anne McClintock, *Imperial Leather: Race, Gender, and Sexuality in the Colonial Contest* (Routledge, 1995), 47.

78. Chelmsford Report, 15.

79. See L. Lindop, Surgeon, HMS *Algerine,* "Report," February 1, 1911, NLA, MS 1058; Cecil Hunter Rodwell, *Report on a Visit to Pitcairn Island* (HMSO, 1921); both are discussed in chapter 3.

80. Houston, "Interview with Mr. 54, Memo of Conversation Dictated May 19," CO 537/462.

81. Notably cited in Chelmsford Report, 2.

82. Claus Gossler, "The Commercial Career of James Lyle Young: Contemporary Evaluations of a Notable South Seas Character," *Journal of Pacific History* 44, no. 2 (September 2009): 209–220.

83. Letter from J. L. Young to H. A. Richards, July 12, 1913, FO 687/15.

84. Chelmsford Report, 2.

85. Scott, *Domination and the Arts of Resistance.*

86. Houston, "Visit of H. E. the Governor and the Deputy Administrator," paragraphs 1899–1900, CO 537/462.

87. Houston, "Visit of H. E. the Governor and the Deputy Administrator," paragraph 1901, CO 537/462.

Chapter 3: Taking Measure

1. "A Racial Laboratory," *New York Times*, February 28, 1935, 18.

2. "A Racial Laboratory."

3. Nordhoff and Hall, *Mutiny on the Bounty;* Nordhoff and Hall, *Men against the Sea;* Nordhoff and Hall, *Pitcairn's Island; Mutiny on the Bounty* (Metro-Goldwyn-Mayer, 1936).

4. Stepan, *Idea of Race in Science*, especially chapter 5.

5. Adrian Young, "Putting Anthropology in Its (Hospitable) Place: Harry Shapiro's Fieldwork on Pitcairn Island, 1934–1935," *History and Anthropology* 33, no. 5 (October 20, 2022): 631–651, https://doi.org/10.1080/02757206.2020.1762591.

6. In many accounts, science self-corrects; see Gould, *The Mismeasure of Man;* Pat Shipman, *The Evolution of Racism: Human Differences and the Use and Abuse of Science* (Harvard University Press, 1994); Paul Farber, "Race-Mixing and Science in the United States," *Endeavour* 27, no. 4 (December 2003): 166–170; for a detailed history of physical anthropology or race science in its contexts, see Frank Spencer, *Ecce Homo: An Annotated Bibliographic History of Physical Anthropology* (Greenwood Press, 1986); Michael A. Little and Kenneth A. R. Kennedy, eds., *Histories of American Physical Anthropology in the Twentieth Century* (Lexington, 2009); for early external histories, see Stepan, *The Idea of Race in Science;* Elazar Barkan, *The Retreat of Scientific Racism: Changing Concepts of Race in Britain and the United States between the World Wars* (Cambridge University Press, 1993); Alice L. Conklin, *In the Museum of Man: Race, Anthropology, and Empire in France, 1850–1950* (Cornell University Press, 2013).

7. Sebastián Gil-Riaño, "Relocating Anti-Racist Science: The 1950 UNESCO Statement on Race and Economic Development in the Global South," *British Journal for the History of Science* 51, no. 2 (June 2018): 281–303, https://doi.org/10.1017/S0007087418000286; Susan Lindee and Ricardo Ventura Santos, "The Biological Anthropology of Living Human Populations: World Histories, National Styles, and International Networks: An Introduction to Supplement 5," *Current Anthropology* 53, no. S5 (April 1, 2012): S3–S16, along with other contributions in that volume. On the partial replacement of race with "population," see Lisa Ganett, "Racism and Human Genome Diversity Research: The Ethical Limits of Population Thinking," *Philosophy of Science* 68, no. 3 (September 2001): 479–487.

8. Tracy Teslow, *Constructing Race: The Science of Bodies and Cultures in American Anthropology* (Cambridge University Press, 2014); Warwick Anderson, "Hybridity, Race, and Science: The Voyage of the *Zaca*, 1934–1935," *Isis* 103, no. 2 (June 1, 2012): 229–253; Anderson, "Racial Hybridity, Physical Anthropology, and Human Biology"; Warwick Anderson and M. Susan Lindee, "Pacific Biologies: How Humans Become Genetic," *Historical Studies in the Natural Sciences* 50, no. 5 (November 2020): 483–497, https://doi.org/10.1525/hsns.2020.50.5.483. Quote from Anderson, "Hybridity, Race, and Science," 253.

9. Shapiro, *The Heritage of the Bounty*, x–xii.

10. Harry L. Shapiro, "A Study of Race Mixture as Exemplified in the Descendants of Tahitian and English Mutineers of the Bounty" (PhD diss., Harvard University, 1926). The results were published as Harry L. Shapiro, *Descendants of the Mutineers of the Bounty*, vol. 11, no. 1 of *Memoirs of the Bernice P. Bishop Museum* (Bishop Museum Press, 1929), 1–106.

11. Eugene Giles, "Two Faces of Earnest A. Hooton," *American Journal of Physical Anthropology* 149, no. S55 (January 1, 2012): 105–113.

12. See especially Regna Darnell's work: Regna Darnell, "The Professionalization of American Anthropology: A Case Study in the Sociology of Knowledge," *Social Science Information* 10, no. 2 (April 1, 1971): 83–103; Regna Darnell, *And Along Came Boas: Continuity and Revolution in Americanist Anthropology* (John Benjamins Publishing, 1998); Regna Darnell, *Invisible Genealogies: A History of Americanist Anthropology* (University of Nebraska Press, 2001). For a popular account, see Charles King, *Gods of the Upper Air: How a Circle of Renegade*

Anthropologists Reinvented Race, Sex, and Gender in the Twentieth Century (Knopf Doubleday, 2020).

13. Among them are Joseph Birdsell, Sherwood Washburn, William Howells, Stanley Garn, and Carleton Coon.

14. Franz Boas, "Changes in the Bodily Form of Descendants of Immigrants," *American Anthropologist* 14, no. 3 (July 9, 1912): 530–562; Franz Boas, "Race Problems in America Today," *Science* 29, no. 752 (May 28, 1909): 839–849.

15. Rachel Caspari, "1918: Three Perspectives on Race and Human Variation," *American Journal of Physical Anthropology* 139, no. 1 (2009): 5–15, https://doi.org/10.1002/ajpa.20975.

16. Arthur Keith, "The Physical Characteristics of Two Pitcairn Islanders," *Man* 17, no. 88 (August 1917): 121–131, https://doi.org/10.2307/2788792.

17. Eugen Fischer, *Die Rehebother Bastards und das Bastardierungsproblem beim Menschen* [The Rehebother basters and the race mixture problem in humans] (Gustav Fischer, 1913). For historical accounts, see Benjamin Madley, "From Africa to Auschwitz: How German South West Africa Incubated Ideas and Methods Adopted and Developed by Nazis in Eastern Europe," *European History Quarterly* 35, no. 3 (July 2005): 429–464; Niels C. Lösch, *Rasse als Konstrukt: Leben und Werk Eugen Fischers* [Race as construct: The life and work of Eugen Fischer] (Lang, 1997); Bernhard Gessler, *Eugen Fischer, 1874–1967: Leben und Werk des Freiburger Anatomen, Anthropologen und Rassenhygienikers bis 1927* [Eugen Fischer: 1874–1967: The life and work of the Freiburg anatomist, anthropologist, and race hygenicist until 1927] (Lang, 2000); Heiner Fangerau, "Rassenhygiene und Öffentlichkeiten: Die Popularisierung des rassenhygienischen Werkes von Erwin Baur, Eugen Fischer und Fritz Lenz" [Racial hygiene and the public: The popularization of racial hygiene works by Erwin Baur, Eugen Fischer and Fritz Lenz], *Bevölkerungsfragen: Prozesse des Wissenschaftstransfers in Deutschland und Frankreich 1870–1939* [Population questions: Processes of knowledge transfers in Germany and France], ed. Patrick Krassnitzer and Petra Overath (Bölau, 2007), 131–154.

18. Henrice Altnik, "An American Race Laboratory: Jamaica, 1865–1940," *Wadabagei: A Journal of the Caribbean and Its Diaspora* 10, no. 3 (Fall 2007): 53–83; Charles B. Davenport and Morris Steggerda, *Race Crossing in Jamaica* (Carnegie Institution, 1929).

19. Charles B. Davenport, "The Effects of Race Intermingling," *Proceedings of the American Philosophical Society* 56, no. 4 (1917): 367.

20. William E. Castle, *Genetics and Eugenics: A Textbook for Students of Biology and a Reference Book for Plant and Animal Breeders* (Harvard University Press, 1921), 68–69.

21. Harry Shapiro, *Norfolk Diary*, Shapiro Papers, American Museum of Natural History (hereafter AMNH), box 33.

22. Shapiro, *Descendants*, 3.

23. "Pioneer of Norfolk Island," *The Mercury*, March 19, 1934.

24. "Sport and Natural History in Norfolk Island," *Country Life*, April 11, 1903, 486.

25. Harry L. Shapiro, "The Romance of the Norfolk Islanders: Modern Descendants of the Mutineers Furnish Material for Research," *Scientific American* 135, no. 3 (September 1926): 182.

26. J. W. Parnell, *Norfolk Island: Report of the Administrator* (Government of Australia, 1923), 5–8.

27. Shapiro, *Descendants*, 33.

28. Harry Shapiro Papers, AMNH, box 33.

29. Shapiro, *Descendants*, 32.

30. Shapiro, *Descendants*, 32.

31. Amos Morris-Reich, "Anthropology, Standardization and Measurement: Rudolf Martin and Anthropometric Photography," *British Journal for the History of Science* 46, no. 3 (September 2013): 487–516.

32. Shapiro, *Descendants*, 33–34, 48.

33. Earnest A. Hooton, "Progress in the Study of Race Mixtures, with Special Reference to Work Carried On at Harvard University," *Proceedings of the American Philosophical Society* 65 (1926): 319.

34. Shapiro, *Descendants*, 1–8.

35. Letter from Herbert Gregory to Harry Shapiro, March 10, 1927, Shapiro Papers, box 4, folder 16.

36. Letter from Emory to Shapiro, April 9, 1927, Shapiro Papers, box 4, folder 16.

37. See letters from Shapiro to Gregory, December 8, 1927; March 24, 1928; January 11, 1929; and letters from Gregory to Shapiro, January 11, 1928; November 17, 1929, Shapiro Papers, box 4, folder 16.

38. Shapiro, *Descendants*, 9.

39. Shapiro, *Descendants*, 32.

40. Shapiro, *Descendants*, 32.

41. Shapiro, *Descendants*, 56.

42. Shapiro, *Descendants*, 41.

43. Shapiro, *Descendants*, 33.

44. Shapiro, *Descendants*, 69.

45. Harry L. Shapiro, "Results of Inbreeding on Norfolk Island," *Science* 65, no. 1693 (1927): 10.

46. Melville J. Herskovits, "A Critical Discussion of the 'Mulatto Hypothesis,'" *Journal of Negro Education* 3, no. 3 (July 1, 1934): 389–402.

47. Letter from R. R. Gates to Shapiro, September 25, 1930, Shapiro Papers, box 3.

48. J. C. Trevor, "Some Anthropological Characteristics of Hybrid Populations," *Eugenics Review* 30, no. 1 (April 1938): 21–31; K. Wagner, "The Variability of Hybrid Populations," *American Journal of Physical Anthropology* 16, no. 3 (January 1, 1932): 283–307.

49. Anthony M. Ludovici, "Eugenics and Consanguineous Marriages," *Eugenics Review* 25, no. 3 (October 1933): 147–155.

50. Shapiro, "The Romance of the Norfolk Islanders," 182.

51. Hooton, "Progress in the Study of Race Mixtures," 319.

52. L. C. Dunn, "An Anthropometric Study of Hawaiians of Pure and Mixed Blood, Based upon Data Collected by Alfred M. Tozzer," *Papers of the Peabody Museum of American Archaeology and Ethnology, Harvard University* 11, no. 3 (1928): 95. See also Christine Leah Manganaro, "Assimilating Hawai'i: Racial Science in a Colonial 'Laboratory,' 1919–1939" (PhD diss., University of Minnesota, 2012), http://conservancy.umn.edu/handle/11299/134014.

53. Shapiro, *Heritage of the Bounty*, xiv.

54. Templeton Crocker, *The Cruise of the Zaca* (Harper, 1933); "Scientific Survey of South-Eastern Polynesia," *Nature* 133, no. 3362 (April 7, 1934): 527.

55. Templeton Crocker, Draft of "*Zaca* Sails South: A Non-Scientific Diary," 1–3, Crocker Papers, California Academy of Sciences, box 2, folder 7–8 (hereafter "*Zaca* Sails South"). The Crocker Papers contain different versions of Crocker's journal across several boxes and folders; I will make note when drawing from other iterations.

56. Letter from (unknown, presumed to be Roy Chapman Andrews) to Frank Chapman, March 22, 1934, Central Archives, AMNH, 1216 1934.

57. Letter from Robert Cushman Murphy to George H. Sherwood, January 4, 1936, Central Archives, AMNH, 1216 1934; letter from Roy Chapman Andrews to Leonard Sanford, February 23, 1934, Central Archives, AMNH, 1216 1934.

58. Letter from Templeton Crocker to Henry Fairfield Osborn, February 7, 1933, Central Archives, AMNH, 1216.

59. Crocker, "*Zaca* Sails South," 20; Harry Shapiro, *On Board the Zaca, September 15, 1934–January 31, 1935*, Shapiro Papers, box 33 (hereafter *On Board the Zaca*). The latter are

Shapiro's typed field notes. His papers also include different, handwritten notes: Harry Shapiro, *Pitcairn Notebook* (December 1934), Shapiro Papers, box 146.

60. Letter to Kenneth Emory from Harry Shapiro, February 6, 1934, Shapiro Papers, box 3, folder 17.

61. Shapiro, *On Board the Zaca*, 1.

62. Shapiro, *On Board the Zaca*, 9. For more on Shapiro and Melville, see Anderson, "Hybridity, Race, and Science."

63. Shapiro, *On Board the Zaca*, 43.

64. Shapiro, *On Board the Zaca*, 57.

65. Shapiro, *On Board the Zaca*, 38.

66. Crocker, "*Zaca* Sails South," 89; Shapiro, *On Board the Zaca*, 20, 37.

67. Crocker, "Manuscript Eastern Polynesia," 102, Crocker Papers, box 2, folder 11. This is another draft of his *Zaca* journey, with revisions (hereafter "Eastern Polynesia").

68. Shapiro, *Heritage of the Bounty*, 5.

69. Shapiro, *On Board the Zaca*, 131.

70. Shapiro, *On Board the Zaca*, 137.

71. Shapiro, *On Board the Zaca*, 132.

72. Shapiro, *Heritage of the Bounty*, 7.

73. Shapiro, *On Board the Zaca*, 133; Shapiro, *Heritage of the Bounty*, 9.

74. Shapiro, *Heritage of the Bounty*, 7, 10, 14.

75. Shapiro, *On Board the Zaca*, 151–152.

76. Shapiro, *On Board the Zaca*, 135–137.

77. Shapiro, *Heritage of the Bounty*, 17.

78. Shapiro, *Heritage of the Bounty*, 24–25.

79. Shapiro, *On Board the Zaca*, 138–139; Shapiro, *Heritage of the Bounty*, 25–26.

80. Letter from H. Bury Palliser to the Secretary of the Admiralty, April 25, 1898, enclosure no. 1, *Pitcairn Island: Correspondence Relating to the Condition of the Pitcairn Islanders* (HMSO, 1899), 1.

81. Letter from Henry Dyke to the Secretary of the Admiralty, November 23, 1897, enclosure no. 1, *Pitcairn Island: Correspondence Relating to the Condition of the Pitcairn Islanders*, 2.

82. Rodwell, *Report on a Visit to Pitcairn Island*, 4, 8.

83. D. Colqohoun, "Impressions of the Inhabitants of the Island," Annexure 2 to Rodwell, *Report on a Visit to Pitcairn Island*, 4. His report was excerpted for the scientific community in D. Colqohoun, "Notes," *Nature* 109, no. 2727 (February 2, 1922): 150.

84. H. G. Pilling, *Report on a Visit to Pitcairn Island* (HMSO, 1930), 6.

85. Crocker, "*Zaca* Sails South," 96.

86. Crocker, "Eastern Polynesia," 107.

87. Shapiro, *On Board the Zaca*, 159.

88. Shapiro, *On Board the Zaca*, 161.

89. Crocker, "Eastern Polynesia," 112.

90. Crocker, "*Zaca* Sails South," 99; three small notebooks contain epitaphs from the cemetery, in Crocker Papers, box 6, folder 12.

91. James Norman Hall, *Tale of a Shipwreck* (Houghton Mifflin, 1934), 62–64.

92. Shapiro, *Heritage of the Bounty*, 120.

93. Shapiro, *Heritage of the Bounty*, 22–23.

94. Shapiro, *On Board the Zaca*, 161.

95. Shapiro, *On Board the Zaca*, 149.

96. For more on secrecy and anthropology, see Tony Crook, *Anthropological Knowledge, Secrecy and Bolivip, Papua New Guinea: Exchanging Skin* (Oxford University Press, 2007); A. Johnson, "Secretsharers."

97. Crocker, "Eastern Polynesia," 113.

98. Crocker, "Eastern Polynesia," 107, 114.

99. Harry Shapiro, "The Pitcairn Islanders" (unpublished lecture, n.d.), Shapiro Papers, box "Pitcairn Island Work—Press Clippings."

100. Shapiro, *Heritage of the Bounty*.

101. Letter from Shapiro to Burley Warren, June 1936, Shapiro Papers, box 10.

102. Margaret Mead, *Coming of Age in Samoa: A Psychological Study of Primitive Youth for Western Civilization* (Morrow, 1928); Nancy Lutkehaus, *Margaret Mead: The Making of an American Icon* (Princeton University Press, 2008); Peter Mandler, *Return from the Natives* (Yale University Press, 2013).

103. Letter from Shapiro to Hooton, March 12, 1935, Shapiro Papers, box 5, folder 2.

104. Shapiro, *Heritage of the Bounty*, 197–198.

105. Shapiro, *Heritage of the Bounty*, 231–232.

106. Shapiro, *Heritage of the Bounty*, 196.

107. Shapiro, *Heritage of the Bounty*, 254.

108. Letter from Franz Boas to Shapiro, October 26, 1937, Shapiro Papers, box 1, folder 23.

109. Letter from Shapiro to Hooton, October 22, 1935, Shapiro Papers, box 5, folder 2.

110. Letter from Hooton's office to Shapiro, October 22, 1935, Shapiro Papers, box 5, folder 2; letter from Hooton to Shapiro, October 30, 1935, Shapiro Papers, box 5, folder 2.

111. Earnest A. Hooton, "Plain Statements about Race," *Science* 83, no. 2161 (May 29, 1936): 511–513.

112. Emil Rogner, *Pitcairn-Island: Ein Beitrag zur Lehre von der Entfaltung Menschlicher Gesellungen* [Pitcairn Island: A contribution to the study of the development of human society], Nürnberger Beiträge zu den Wirtschafts- und Sozialwissenschaften [Nuremberg contributions to economics and social science] 78 (Verlag der Hochschulbuchhandlung, Krisdie, 1940).

113. Rogner, *Pitcairn-Island*, 58.

114. Ruth Benedict, *Race: Science and Politics* (Viking, 1945), 43–44, 52.

115. Ashley Montagu, *Man's Most Dangerous Myth: The Fallacy of Race* (Columbia University Press, 1945), 110–111; M. F. Ashley Montagu, "The Creative Power of Ethnic Mixture," *Psychiatry* 5, no. 4 (November 1, 1942): 523–536.

116. Franz Boas, *Race, Language, and Culture* (University of Chicago Press, 1940), 174.

117. Letter from Boas to Shapiro, March 30, 1942, Shapiro Papers, box 1, folder 23; Franz Boas, "Individual, Family, Population, and Race," *Proceedings of the American Philosophical Society* 87, no. 2 (August 16, 1943): 161–164.

118. Cf. Jerold D. Cummins and John L. Kane Jr., "Miscegenation, the Constitution, and Science," *Dicta* 38 (1961): 24; Stuart W. Conner, "Constitutionality of Miscegenetic Marriages," *Montana Law Review* 11 (1950): 53; John P. Jackson, *Science for Segregation: Race, Law, and the Case against Brown v. Board of Education* (New York University Press, 2005).

119. H. L. Shapiro, "Some Reflections of an Anthropologist on the Future of Our Population," *Proceedings of the American Philosophical Society* 80, no. 4 (February 15, 1939): 587–600.

120. Sherwood Washburn, "Section of Anthropology: The New Physical Anthropology," *Transactions of the New York Academy of Sciences* 13, no. 7 (1951): 298–304. For a classic treatment of Washburn's project and its gendered Cold War context, see Donna J. Haraway, *Primate Visions: Gender, Race, and Nature in the World of Modern Science*, repr. ed. (Routledge, 1990). For a more recent take, see Jessica Mikels-Carrasco, "Sherwood Washburn's New Physical Anthropology: Rejecting the 'Religion of Taxonomy,'" *History and Philosophy of the Life Sciences* 34, no. 1–2 (2012): 79–101.

121. "Hall of Human Biology," Shapiro Papers, box 83, folder "Anthropology Notes, HLS Miscellaneous Topics."

122. President of the American Ethnological Society 1942–1943, president of the American Eugenics Society 1955–1962, and president of the American Anthropological Association in 1948.

123. Letter from Osborne to members of the Eugenics Research Association, October 25, 1938, Shapiro Papers, box 12.

124. Frank B. Livingstone, "Genetics, Ecology and the Origins of Incest and Exogamy," *Current Anthropology* 10, no. 1 (February 1969): 59.

125. See Carleton Coon, *The Races of Europe* (Macmillan, 1939); Carleton Coon, *The Origin of Races* (Alfred A. Knopf, 1963); John P. Jackson Jr., "'In Ways Unacademical': The Reception of Carleton S. Coon's *The Origin of Races*," *Journal of the History of Biology* 34, no. 2 (June 1, 2001): 247–285; Peter Collopy, "Race Relationships: Collegiality and Demarcation in Physical Anthropology," *Journal of the History of the Behavioral Sciences* 51, no. 3 (Summer 2015): 237–260. For the turn to "population thinking," see Michael A. Little, "Human Population Biology in the Second Half of the Twentieth Century," *Current Anthropology* 53, no. S5 (2012): S126–S138, https://doi.org/10.1086/662440; Jonathan Marks, "The Origins of Anthropological Genetics," *Current Anthropology* 53, no. S5 (April 1, 2012): S161–S172, https://doi.org/10.1086/662333; Jenny Bangham and Soraya de Chadarevian, "Human Heredity after 1945: Moving Populations Centre Stage," *Studies in History and Philosophy of Science Part C: Studies in History and Philosophy of Biological and Biomedical Sciences* 47, Part A (September 1, 2014): 45–49, https://doi.org/10.1016/j.shpsc.2014.05.005.

126. Stanley M. Garn and Carleton S. Coon, "On the Number of Races of Mankind," *American Anthropologist* 57, no. 5 (October 1955): 999; see also the responses of both anthropologists to Andrzej Wiercinski and Tadeusz Bielicki, "The Racial Analysis of Human Populations in Relation to Their Ethnogenesis," *Current Anthropology* 3, no. 1 (February 1962): 2, 9–46. Joined by Birdsell, they also defended race as a category in Carleton S. Coon et al., *Races: A Study of the Problems of Race Formation in Man* (Charles Thomas, 1950).

127. See also lingering eugenic interest in the Pitcairn case, as in Macfarlane Burnet, "Migration and Race from the Genetic Angle," *Eugenics Review* 51, no. 2 (July 1959): 93–97.

128. Harry L. Shapiro, "Forensic Anthropology," *Annals of the New York Academy of Sciences*, 318 (December 1978): 3–9.

129. For historical treatments, see Michelle Brattain, "Race, Racism, and Antiracism: UNESCO and the Politics of Presenting Science to the Postwar Public," *American Historical Review* 112, no. 5 (December 1, 2007): 1386–1413, https://doi.org/10.1086/ahr.112.5.1386; Anthony Q. Hazard, "A Racialized Deconstruction? Ashley Montagu and the 1950 UNESCO Statement on Race," *Transforming Anthropology* 19, no. 2 (October 1, 2011): 174–186; Gil-Riaño, "Relocating Anti-Racist Science."

130. *The Race Concept: Results of an Inquiry* (UNESCO, 1952); Harry L. Shapiro, "Revised Version of UNESCO Statement on Race," *American Journal of Physical Anthropology* 10, no. 3 (September 3, 1952): 335–336.

131. Shapiro, "Revised Version of UNESCO Statement on Race."

132. Harry Shapiro, "Race Mixture," in *The Race Question in Modern Science* (Columbia University Press, 1961), 343–392.

133. Shapiro, *Heritage of the Bounty*, 137–138, 217.

134. "A Racial Laboratory," 18; see also "Pitcairn Refutes Inbreeding Harm: Biologist's Study of Island Records Shows Mental and Physical Vigor Unimpaired," *New York Times*, November 7, 1936, N11.

135. Kohler, *Landscapes and Labscapes*, 3.

136. Shapiro, *Heritage of the Bounty*, 193.

137. Shapiro, *The Heritage of the Bounty*; Harry Shapiro, *The Heritage of the Bounty*, rev. with new postscript (Anchor Books, 1962); Harry Shapiro, *The Pitcairn Islanders (Formerly the Heritage of the Bounty)* (Simon and Schuster, 1968); Harry Shapiro, *The Heritage of the Bounty* (AMS Press, 1979); Harry Shapiro, *The Pitcairn Islanders*, rev. ed. (Simon and Schuster, 1986).

138. Cf. John Phillip Gillian, *The Ways of Men: An Introduction to Anthropology* (Appleton-Century-Crofts, 1948), 132; Juan Comos, *Manual of Physical Anthropology* (Charles Thomas,

1960), 170–171; Gabriel Lasker, *The Evolution of Man: A Brief Introduction to Physical Anthropology* (Holt, 1961), 177–179; Alan Robin Beals et al., *Culture in Process* (Holt, Rinehart and Winston, 1967), 23–24; Gretal Pelto and Pertti Pelto, *The Human Adventure: An Introduction to Anthropology* (Macmillan, 1976), 109; Robert Jurmain et al., *Introduction to Physical Anthropology* (Wadsworth, 2011), 448.

139. Sally Falk Moore, "Part of the Story: A Memoir," *Ethnos* 70, no. 4 (December 1, 2005): 538–566.

140. Adria L. Imada, *An Archive of Skin, an Archive of Kin: Disability and Life-Making during Medical Incarceration* (University of California Press, 2022), 13–16.

Chapter 4: Legitimating Pasts

1. "Pitcairn's Island—The Bounty's Crew," in *Colburn's United Service Magazine and Naval and Military Journal* (Henry Colburn and Richard Bentley, 1829), 2:589–593.

2. Pauline Reynolds, "The Forgotten Women of the 'Bounty' and Their Material Heritage," in *Conference Proceedings: Maori and Pacific Textile Symposium: Whatu Raranga a Kiwa, Understanding and Uniting Mori and Pacific Textiles* (Museum of New Zealand Te Papa Tongarewa, 2012), 2.

3. Donald Patrick Albert, "Teehuteatuaonoa aka 'Jenny,' the Most Traveled Woman on the Bounty: Chronicling Female Agency and Island Movements with Google Earth," *Island Studies Journal* 16, no. 1 (May 2021): 190–208, https://doi.org/10.24043/isj.153.

4. Teehuteatuaonoa, "Account of the Mutineers of the Ship Bounty, and Their Descendants at Pitcairn's Island," *Sydney Gazette and New South Wales Advertiser,* July 17, 1819, 3. http://nla.gov.au/nla.news-article2178824.

5. From a version of Teehuteatuaonoa's account first printed in the *Bengal Hurkaru,* October 2, 1826; reprinted as "Pitcairn's Island—the Bounty's Crew."

6. John Sekora, "Black Message/White Envelope: Genre, Authenticity, and Authority in the Antebellum Slave Narrative," *Callaloo,* no. 32 (1987): 482, https://doi.org/10.2307/2930465.

7. Teehuteatuaonoa, "Account of the Mutineers of the Ship Bounty," 589.

8. For a similar account of maritime translation and adjudication of authority, see Edward A. Alpers and Matthew S. Hopper, "Speaking for Themselves? Understanding African Freed Slave Testimonies from the Western Indian Ocean, 1850s–1930s," *Journal of Indian Ocean World Studies* 1, no. 1 (September 29, 2017): 60–89, https://doi.org/10.26443/jiows.v1i1.20.

9. Barrow, *The Eventful History of the Mutiny;* Murray, *Pitcairn, the Island, the People, and the Pastor;* Belcher, *The Mutineers of the Bounty;* R. A. Young, *Mutiny of the Bounty and Story of Pitcairn Island;* Henry Evans Maude, *Of Islands and Men* (Oxford University Press, 1968); Dening, *Mr Bligh's Bad Language.*

10. Jeannette Mageo and Bruce Knauft, eds., *Authenticity and Authorship in Pacific Island Encounters: New Lives of Old Imaginaries* (Berghahn Books, 2021).

11. "A History of Pitcairn Island," *Signs of the Times* 20, no. 54 (November 19, 1894): 864.

12. Pierre-Philippe Fraiture, *Past Imperfect: Time and African Decolonization, 1945–1960* (Oxford University Press, 2021); Priya Satia, *Time's Monster* (Harvard University Press, 2020); Ian Hesketh, *The Science of History in Victorian Britain: Making the Past Speak,* Science and Culture in the 19th Century 12 (Routledge, 2015); Rolf Torstendahl, *The Rise and Propagation of Historical Professionalism* (Routledge, 2014); Hayden White, *Metahistory: The Historical Imagination in Nineteenth-Century Europe* (1974; repr. Johns Hopkins University Press, 2014).

13. The history of history writing by women, especially in Britain, is vast. See Deirdre David, *Rule Britannia: Women, Empire, and Victorian Writing* (Cornell University Press, 1995); Rohan Amanda Maitzen, *Gender, Genre, and Victorian Historical Writing* (Taylor and Francis,

1998); Bonnie G. Smith, *The Gender of History: Men, Women, and Historical Practice* (Harvard University Press, 2000); Christine L. Krueger, "Why She Lived at the PRO: Mary Anne Everett Green and the Profession of History," *Journal of British Studies* 42, no. 1 (January 2003): 65–90, https://doi.org/10.1086/342686; Devoney Looser, *British Women Writers and the Writing of History, 1670–1820* (Johns Hopkins University Press, 2005); Lynette Felber, ed., *Clio's Daughters: British Women Making History, 1790–1899* (University of Delaware Press, 2007); Elise Garritzen, *Pasha and His Historic Harem: Edward A. Freeman, Edith Thompson and the Gendered Personae of Late-Victorian Historians* (Manchester University Press, 2019).

14. For a reading of Young against American missionary and imperial gazes, see Sebastian Jablonski, "The 'Uncorrupted' Paradise: Religion and Imperial Epistemic Violence on Pitcairn Island," in "New Scholarship in New Zealand and Pacific Studies Part 2," special issue, *Journal of New Zealand and Pacific Studies* 9 (December 1, 2021): 199–214, https://doi.org/10.1386/nzps_00070_1.

15. *The Women of Pitcairn Island*, directed by Jean Yarbough (20th Century Fox, 1956).

16. Gayatri Chakravorty Spivak, "Can the Subaltern Speak?," in *Marxism and the Interpretation of Culture*, ed. Cary Nelson and Lawrence Grossberg (Macmillan, 1988), 271–313.

17. Rosalind C. Morris, *Can the Subaltern Speak? Reflections on the History of an Idea* (Columbia University Press, 2010).

18. Bronwen Douglas and Chris Ballard, "Contact Tracing: The Materiality of Encounters," *History and Anthropology* 33, no. 1 (January 1, 2022): 1–16, https://doi.org/10.1080/02757206.2021.1987234; Nicholas Thomas, *Entangled Objects: Exchange, Material Culture, and Colonialism in the Pacific* (Harvard University Press, 1991).

19. R. A. Young, "Miss Young's Letter to the 'Sunday World,'" in *Mutiny of the Bounty and Story of Pitcairn Island*, 258.

20. Mary S. Daniels, "The Mutineers of the 'Bounty,'" *Methodist Magazine* 36, no. 6 (December 1892): 565.

21. Letter from Henry Walter Petch to R. T. Simons, September 21, 1904, TNA, FO/687/10 Pitcairn Island Correspondence (emphasis in original).

22. Herbert Ford, *Island of Tears: John I. Tay and the Story of Pitcairn* (Pacific Press, 1989).

23. R. A. Young, *Mutiny of the Bounty and Story of Pitcairn Island*, 235–236.

24. "Ein Brief aus Pitcairn" [A letter from Pitcairn], in *Evangelisches Missions-Magazin* (Verlag der Missionsbuchhandlung, 1883), 202–213; letter from Rosalind Amelia Young to Rev. W. B. Keen, May 16, 1878, in Alfred McFarland, *Mutiny in the "Bounty" and Story of the Pitcairn Islanders* (J. J. Moore, 1884), 234–235; Rosalind A. Young, "Letters from Pitcairn Island," *Overland Monthly and Out West Magazine* 18, no. 105 (September 1891): 294–308.

25. R. A. Young, "Letters from Pitcairn Island," 296–297.

26. R. A. Young, "Letters from Pitcairn Island," 296–297.

27. Adrian Young, "Bounty Relics: Trading in the Legacy of Myth and Mutiny," in *The Bounty from the Beach: Cross-Cultural and Cross-Disciplinary Essays*, ed. Sylvie Largeaud-Ortega (Australian National University Press, 2018), 207–234.

28. R. A. Young, "Letters from Pitcairn Island," 287.

29. Rosalind A. Young, "How the Pitcairn Islanders Spend Sunday," *The Advance*, September 14, 1882; Rosalind A. Young, "News from Pitcairn," *Signs of the Times*, November 19, 1906, 572.

30. "A Cruise among the South Sea Islands," in *Leisure Hour; an Illustrated Magazine for Home Reading* (1890), 348–350.

31. R. A. Young, *Mutiny of the Bounty and Story of Pitcairn Island*, 13.

32. White, *Metahistory;* Axel Schneider and Daniel Woolf, *The Oxford History of Historical Writing*, vol. 5, *Historical Writing Since 1945* (Oxford University Press, 2011); Alun Munslow, *A History of History* (Routledge, 2012).

33. The original register left the island during the 1850s, and it is unclear which version she references. R. A. Young, *Mutiny of the Bounty and Story of Pitcairn Island*, 92.

34. Rosalind Amelia Young, "Pitcairn Island, Aug. 26, 1881," *Century Illustrated Monthly Magazine* 24, no. 2 (1882).

35. R. A. Young, *Mutiny of the Bounty and Story of Pitcairn Island*, 203–211.

36. Quoted in R. A. Young, *Mutiny of the Bounty and Story of Pitcairn Island*, 34–35.

37. Elise Garritzen, "Women Historians, Gender and Fashioning the Authoritative Self in Paratexts in Late-Victorian Britain," *Women's History Review* 30, no. 4 (June 7, 2021): 650–668.

38. Hook Milton, "Gates, Edward Harmon (1855–1940)," in *Encyclopedia of Seventh-Day Adventists*, published September 5, 2020, https://encyclopedia.adventist.org/article?id=A9C7.

39. Edward Harmon Gates, "Introduction," in R. A. Young, *Mutiny of the Bounty and Story of Pitcairn Island*, xii.

40. Gates, "Introduction," viii.

41. Gates, "Introduction," vii.

42. R. A. Young, *Mutiny of the Bounty and Story of Pitcairn Island*, 23–24.

43. R. A. Young, *Mutiny of the Bounty and Story of Pitcairn Island*, 30.

44. R. A. Young, *Mutiny of the Bounty and Story of Pitcairn Island*, 31.

45. R. A. Young, *Mutiny of the Bounty and Story of Pitcairn Island*, 65.

46. See Reynolds, "The Forgotten Women of the 'Bounty' and Their Material Heritage."

47. R. A. Young, *Mutiny of the Bounty and Story of Pitcairn Island*, 98.

48. R. A. Young, *Mutiny of the Bounty and Story of Pitcairn Island*, 89.

49. R. A. Young, *Mutiny of the Bounty and Story of Pitcairn Island*, 106, 179.

50. R. A. Young, *Mutiny of the Bounty and Story of Pitcairn Island*, 181.

51. Duncan Bell, *The Idea of Greater Britain: Empire and the Future of World Order, 1860–1900* (Princeton University Press, 2009).

52. Charles Wentworth Dilke, *Greater Britain: A Record of Travel in English-Speaking Countries, during 1866–1867* (Macmillan, 1869), 230.

53. See especially the discussion of the "indolence" trope in chapter 2.

54. R. A. Young, *Mutiny of the Bounty and Story of Pitcairn Island*, 180.

55. R. A. Young, *Mutiny of the Bounty and Story of Pitcairn Island*, 146.

56. R. A. Young, *Mutiny of the Bounty and Story of Pitcairn Island*, 146.

57. Greg Dening wrote that "their myths about 'blacks' were unequivocal: they were treacherous, inferior, dangerous. . . . Part of their innocence, visitors discovered, was the disarming frankness with which they discussed the evils of being 'black.'" Dening, *Mr Bligh's Bad Language*, 336.

58. Daniel Pick, *Faces of Degeneration: A European Disorder, 1848–1918* (Cambridge University Press, 1989).

59. Daniels, "The Mutineers of the 'Bounty,'" 565.

60. R. A. Young, "Letters from Pitcairn Island," 305–306.

61. R. A. Young, *Mutiny of the Bounty and Story of Pitcairn Island*, 253.

62. "Lesson 33: Story of Pitcairn Island," *Youth's Instructor*, May 23, 1911; "A Word for the Reading Course," *North Pacific Union Gleaner*, September 21, 1910.

63. Letter from A. H. Young to David Nield, March 29, 1929, Reports and Statements Regarding the Pitcairn Group, part 1, A—Pitcairn Island, box 1, Papers of Henry Evans Maude, University of Adelaide Special Collections.

64. Letter from David Nield to the high commissioner for the Western Pacific, June 14, 1928, Reports and Statements Regarding the Pitcairn Group, part 1, A—Pitcairn Island, box 1, Papers of Henry Evans Maude, University of Adelaide Special Collections.

65. "Prominent Pitcairners—Part 5, Rosalind Amelia Young," accessed September 1, 2021, http://www.stamps.gov.pn/RosalindAmeliaYoung.html.

66. Henry Evans Maude at least listed her account as a "standard work," but did not rate her as a "classic," and suggested she added little to historians' knowledge of the island. Maude, *Of Islands and Men*, footnote 1. Greg Dening, perhaps the most famous of the *Bounty* and Pitcairn historians, credited her for giving "an insider's view gently, piously." Dening, *Mr Bligh's Bad Language*, 395.

67. H. E. Maude, "Pacific History—Past, Present and Future," *Journal of Pacific History* 6, no. 1 (January 1971): 24, https://doi.org/10.1080/00223347108572180.

68. Maude, *Of Islands and Men*.

69. For instance, famously, Hugh Trevor-Roper, *The Rise of Christian Europe* (Thames and Hudson, 1965).

70. Satia, *Time's Monster*.

71. For a biography of Maude, see Susan Woodburn, *Where Our Hearts Still Lie: Harry and Honor Maude in the Pacific Islands* (Crawford House Publishing Australia, 2003), as well as his obituary by Niel Gunson, "Harry Maude: Unimane, Statesman and Pacific Historian," *Journal of Pacific History* 42, no. 1 (2007): 109–118.

72. Robert Langdon, "Harry Maude: Shy Proconsul, Dedicated Pacific Historian," in *The Changing Pacific: Essays in Honour of H. E. Maude*, ed. Niel Gunson (Oxford University Press, 1978), 1–21.

73. Maude, *Of Islands and Men*, xii.

74. Maude, *Of Islands and Men*, xiii.

75. H. C. Maude and H. E. Maude, "Adoption in the Gilbert Islands," *Journal of the Polynesian Society* 40, no. 4(160) (1931): 225–235; H. C. Maude and H. E. Maude, "The Social Organization of Banaba or Ocean Island, Central Pacific," *Journal of the Polynesian Society* 41, no. 4(164) (1932): 262–301; H. C. Maude and H. E. Maude, "String-Figures from the Gilbert Islands," *Journal of the Polynesian Society* 45, no. 2(178) (1936): 1–16.

76. Maude, "The Colonization of the Phoenix Islands," in *Of Islands and Men*, 330–331.

77. H. E. Maude, "Memorandum. Inception of the scheme for colonizing," February 12, 1942, 3, Harry Evans and Honor Maude Papers, University of Adelaide Rare Books and Manuscript Collections (hereafter Maude Papers), part I, series E, folder 1: Reports and Correspondence Leading to the Colonization of the Phoenix Islands, part I.

78. H. E. Maude, *Colonization of the Phoenix Islands by the Surplus Population of the Gilbert and Ellice Islands* (Government Press, 1937), 3–4.

79. Letter from R. H. Garvey to H. E. Maude, November 25, 1938; telegram from the secretary of state to the high commissioner, June 2, 1938; memorandum by A. F. Richards (high commissioner), February 14, 1938; Maude, *Colonization of the Phoenix Islands*, 14.

80. Maude, *The Colonization of the Phoenix Islands*, 341.

81. Letter from Maude to high commissioner, January 10, 1945, Maude Papers, part I, series J, section 18, 1945.

82. Nicholas Hoare, "Re-mining Makatea: People, Politics and Phosphate Rock" (PhD diss., Australian National University, 2020), https://www.proquest.com/docview/2564078845/abstract/F25C5F2FEE7E4EF3PQ/1; Katerina Martina Teaiwa, *Consuming Ocean Island: Stories of People and Phosphate from Banaba* (Indiana University Press, 2014); Katerina Teaiwa, "Ruining Pacific Islands: Australia's Phosphate Imperialism," *Australian Historical Studies* 46, no. 3 (September 2, 2015): 374–391, https://doi.org/10.1080/1031461X.2015.1082609.

83. H. E. Maude, "The Future of the Banaban Population of Ocean Island; with Special Relation to Their Lands and Funds," memorandum, September 2, 1946, Maude Papers, series F9, Memorandum to Secretary, 1946.

84. See Katerina Martina Teaiwa, "Our Sea of Phosphate: The Diaspora of Ocean Island," in *Indigenous Diasporas and Dislocations*, ed. Charles D. Thompson and Graham Harvey (Routledge, 2017), 169–192; Jane McAdam, "Historical Cross-Border Relocations in the Pacific: Lessons for Planned Relocations in the Context of Climate Change," *Journal of Pacific*

History 49, no. 3 (July 3, 2014): 301–327, https://doi.org/10.1080/00223344.2014.953317; Jane McAdam, "Self-Determination and Self-Governance for Communities Relocated across International Borders: The Quest for Banaban Independence," *International Journal on Minority and Group Rights* 24, no. 4 (November 3, 2017): 428–466, https://doi.org/10.1163/15718115-02404001; Julia Edwards, "Phosphate and Forced Relocation: An Assessment of the Resettlement of the Banabans to Northern Fiji in 1945," *Journal of Imperial and Commonwealth History* 41, no. 5 (December 1, 2013): 783–803, https://doi.org/10.1080/03086534.2013.814252.

85. H. E. Maude, "The Purchase of Rabi Island for the Banabans" (n.d.), Maude Papers, series F, folder 11, Undated Paper by Maude.

86. Maude, "The Future of the Banaban Population of Ocean Island," 15.

87. Maude, "The Future of the Banaban Population of Ocean Island," 17.

88. Letter from Maude to Major Holland, July 4, 1946, Maude Papers, part I, series J, section 20, 1947 part I.

89. Maude, "The Future of the Banaban Population of Ocean Island," 28.

90. H. E. Maude, "A General Report Based on Seven Months Residence in the Island during 1940–41; with Suggestions for the Future Welfare of the Community," Maude Papers, part I, series J, section 14, 1941.

91. Letter from Honor Maude to Lady Maude, August 1940, Maude Papers, part I, series J, section 29: Letters Home 1940–46; letter from F. Ward to Maude, June 20, 1940, Maude Papers, part I, series A, School Affairs.

92. Letter from Honor Maude to Lady Maude, August 1940.

93. Maude, "A General Report," 3.

94. Maude, "A General Report," 2.

95. Maude, "A General Report," 6.

96. Maude, "A General Report," 9.

97. Letter from Maude, December 15, 1940, Maude Papers, part I, series A, Instructions for the Guidance of the Local Government.

98. Letter from Maude to Harry Shapiro, October 15, 1940, Maude Papers, part I, series J, section 13, 1940.

99. Maude, "A General Report," 7.

100. Letter from Honor Maude to Lady Maude, March 22, 1940, Maude Papers, part I, series J, section 29: Letters Home 1940–46.

101. Letter from Honor Maude to Lady Maude, April 17, 1940, 7, Maude Papers, part I, series J, section 29: Letters Home 1940–46.

102. Maude, "The Future of the Banaban Population of Ocean Island," 12.

103. Maude, abstract from "Colonization Experiments in the Central Pacific," Seventh Pacific Science Congress, New Zealand, 1949, Maude Papers, part I, series J, section 23, 1949.

104. Letter from James Wightman Davidson to Maude, July 17, 1952; letter from registrar to Maude, April 8, 1952; and letter from Maude to Davidson, June 24, 1953, in Maude Papers, part I, series G, Application for Research Fellowship.

105. Gavan Daws, "Texts and Contexts: A First-Person Note," *Journal of Pacific History* 41, no. 2 (September 2006): 250.

106. Letter from Davidson to Maude, June 29, 1953, Maude Papers, part I, series G, Application for Research Fellowship.

107. Letter from Maude to Davidson, August 5, 1957, Maude Papers, part I, series J, section 27: 1957.

108. Letter from Maude to Dorothy Crozier, May 12, 1958, Maude Papers, part I, series J, section 27: 1958 part I.

109. Maude, "In Search of a Home"; Maude, "Tahitian Interlude"; H. E. Maude, "The History of Pitcairn Island," in Ross and Moverley, *The Pitcairnese Language*, 45–101.

110. H. E. Maude, "The Voyage of the Pandora's Tender," *Mariner's Mirror* 50, no. 3 (1964): 217–235; H. E. Maude, "Some Quotations about the Pitcairnese Language," in *The Pitcairnese Language,* edited by Ross and Moverley, 118–120; H. E. Maude, "The Pitcairn Commemorative Stamp Issue, 1961," *Crown Agents Stamp Bulletin,* no. 346 (1960): 20–24.

111. H. E. Maude, "Review of Adams of the Bounty by Erle Wilson," *Journal of the Polynesian Society* 69, no. 2 (1960): 177–178; H. E. Maude, "Review of Pandora's Last Voyage by Geoffrey Rawson," *Journal of the Polynesian Society* 73, no. 2 (1964): 240–241; H. E. Maude, "Review of Who Caused the Mutiny on the Bounty? By Madge Darby," *Journal of Pacific History* 1 (1966): 246–247; H. E. Maude, "Pitcairn Island," in *Encyclopedia Britannica* (Encyclopedia Britannica, 1967), 17:1116; H. E. Maude, "George Hunn Nobbs (1799–1884)," in *Australian Dictionary of Biography* (Melbourne University Press, 1967), 2:288–289.

112. Letter from Maude to John Gutch, March 11, 1957, Maude Papers, part I, series J, section 27: 1957; letter from Reid Cowell to Maude, October 5, 1960, Maude Papers, part I, series A, Pitcairn Island Commemorative Stamp Issue.

113. H. E. Maude, "A Social and Administrative History of Pitcairn Island," Maude Papers, Part II, F/7—Papers Relating to Pitcairn Island; letter from Maude to J. B. Clayton, April 9, 1962, Maude Papers, part I, series J, section 27: 1962 part I.

114. Letter from Maude to Margaret Titcomb, July 27, 1957, Maude Papers, part I, series J, section 27: 1957.

115. Maude, *Of Islands and Men,* xvi.

116. Maude, *Of Islands and Men,* xvi.

117. Curriculum vitae, March 11, 1952, Maude Papers, part I, series G, Application for Research Fellowship.

118. Letter from Maude to Spivey, January 3, 1945, Maude Papers, part I, series J, section 18.

119. Letter from Maude to Davidson, August 5, 1957, Maude Papers, part I, series J, section 27: 1957.

120. Maude, "In Search of a Home," in *Of Islands and Men,* 1.

121. Letter from Maude to H.B.M.'s consul, Honolulu February 7, 1958, Maude Papers, part I, series J, section 27: 1958 part II.

122. Letter from Maude to the librarian, Royal Geographical Society, May 27, 1962, Maude Papers, part I, series J, section 27: 1962 part I; letter from Gavin Daws to Maude, March 22, 1963, Maude Papers, part I, series J, section 27, 1963 part II; letter from Maude to Henderson-Howat, September 14, 1960, Maude Papers, part I, series J, section 27: 1960 part II.

123. Letter from Maude to Shapiro October 7, 1957, Maude Papers, part I, series J, section 27: 1958 part I.

124. Maude, "The History of Pitcairn Island," 51; Maude, "In Search of a Home," in *Of Islands and Men,* 2.

125. Though he did later call her history "a remarkably good one, though now dated" in private correspondence. See letter from Maude to Joanne Dombrowski, May 3, 1962, Maude Papers, part I, series J, section 27: 1962 part I.

126. Maude, "In Search of a Home," in *Of Islands and Men,* 2–3.

127. Maude, "In Search of a Home," in *Of Islands and Men,* 13.

128. Bronwen Douglas and Doug Munro, "Of Islands and Sandalwood: Shineberg, Maude, and the Hidden History of Trade," in *Texts and Contexts: Reflections in Pacific Islands Historiography,* ed. Brij V. Lal and Doug Munro (University of Hawai'i Press, 2005), 140–153.

129. Maude, "Pacific History: Past, Present, Future," 11.

130. Maude, *Of Islands and Men,* xviii.

131. Maude, "Pacific History: Past, Present, Future," 20.

132. Satia, *Time's Monster.*

133. Maude, "Pacific History: Past, Present, Future," 23–24.

134. Letter from Maude to Leslie Grinsell, February 14, 1960, Maude Papers, part I, series J, section 27: 1960 part I.

135. It is no surprise that Greg Dening, who used stage as metaphor, considered Maude his mentor. See Dening, *Mr Bligh's Bad Language,* xi.

136. Maude, "The History of Pitcairn Island," 51.

137. Maude, "The History of Pitcairn Island," 57.

138. Maude, "Tahitian Interlude," in *Of Islands and Men,* 313.

139. Letter from Paul Hasluck to Ronald Garvey, n.d., enclosure VII in memorandum from Maude to C. R. Lambert, June 17, 1956, Maude Papers, part I, series J, section 27: 1956.

140. Letter from Maude to Lambert, August 18, 1955, Maude Papers, part I, series G, Survey and Report Part I.

141. Maude, "Tahitian Interlude," in *Of Islands and Men,* 295.

142. Maude, "Tahitian Interlude," in *Of Islands and Men,* 302.

143. Maude, "Tahitian Interlude," in *Of Islands and Men,* 312.

144. Letter from Maude to Rolf du Rietz, February 9, 1964, Maude Papers, part 1, series J, section 27: 1964 part I.

145. Letter from Maude to Ried Cowell, February 9, 1963, Maude Papers, part I, series J, section 27, 1963 part II.

146. Letter from Maude to Homer Barnett, June 19, 1963, Maude Papers, part I, series G, Comparative Study of Cultural Change.

147. Letter from Maude to Homer Barnett, June 19, 1963.

148. Letter from Maude to Kenneth Knudson, May 6, 1963, Maude Papers, part I, series G, Comparative Study of Cultural Change.

149. Letter from William T. Stuart to Maude, April 24, 1964, Maude Papers, part I, series G, Comparative Study of Cultural Change.

150. Letter from Maude to William T. Stuart, June 29, 1964, Maude Papers, part I, series G, Comparative Study of Cultural Change.

Chapter 5: Entangled Tapes

1. A. C. Grimson, "The Pitcairnese Text," in Ross and Moverley, *The Pitcairnese Language,* 121.

2. The languages have gone by several names, including "Pitcairnese," "Pitkern," and "Norf'k." I am sympathetic to the latter two, as they have been adopted by many Norfolk Islanders, but in this chapter will refer to the languages as the Pitcairn and Norfolk languages in order to separate analytic and actors' categories.

3. John Reinecke, "Pitcairnese and Norfolk," in *A Bibliography of Pidgin and Creole Languages* (University of Hawai'i Press, 1975), 590–592.

4. Particularly in my use of entanglement as a metaphor, I build on Nicholas Thomas, *Entangled Objects: Exchange, Material Culture, and Colonialism in the Pacific* (Harvard University Press, 1991).

5. For internalist accounts, see Dell H. Hymes, *Essays in the History of Linguistic Anthropology* (John Benjamins Publishing, 1983), or E. F. K. Koerner, *Essays in the History of Linguistics* (John Benjamins Publishing, 2004). For linguists' colonial situation, see Joseph Errington, "Colonial Linguistics," *Annual Review of Anthropology* 30 (2001): 19–39.

6. Peter Mühlhäusler, "The College of St. Barnabas on Norfolk Island and Its Languages: An Early Example of Missionary Language Planning," *Language and Communication* 30, no. 4 (October 2010): 225–239, https://doi.org/10.1016/j.langcom.2010.02.001.

7. John E. Reinecke, "Marginal Languages" (PhD diss., Yale University, 1937).

8. John Holm, *An Introduction to Pidgins and Creoles* (Cambridge University Press, 2000), and Michel DeGraff, "Against Creole Exceptionalism," *Language* 79, no. 2 (June 2003): 391–410; Charles Stewart, ed., *Creolization: History, Ethnography, Theory* (Left Coast Press, 2007).

9. Stina Hasse, "I Am Sitting in a Room," *Body, Space and Technology* 11 (January 1, 2012), https://doi.org/10.16995/bst.71.

10. Brian Freeston, "Note on Pitcairn," April 14, 1949, 8, TNA, CO 537/4933.

11. Ross and Moverley, eds., *The Pitcairnese Language*, 7–8; letter from J. W. Hayward to A. S. C. Ross, July 25, 1961, Letters Additional Collection, Cadbury Research Library Special Collections, University of Birmingham (hereafter L.Add), 1609; letter from E. G. Kedgley to A. S. C. Ross, July 31, 1961, L.Add 1612.

12. Colonial Office Despatch, June 16, 1949, TNA, CO 537/4933.

13. Colonial Office Despatch, June 16, 1949, TNA, CO 537/4933.

14. J. B. Claydon, "Report on Pitcairn Island," January 30, 1954, TNA, CO 1036/114.

15. "Education in Pitcairn Island," 2, TNA, CO 1036/739.

16. Claydon "Report on Pitcairn Island," 19.

17. Letter from A. W. Moverley to the Colonial Office, August 23, 1953, TNA, CO 1023/194.

18. Letter from Jane Moverley to R. Garvey, November 18, 1952, TNA, CO 1023/194.

19. Savingram from assistant high commissioner to secretary of state for the colonies, March 19, 1952, TNA, CO 1023/194.

20. Letter from Roy Sanders to H. A. C. Dobbs, July 7, 1952, TNA, CO 1009/811.

21. Roy Sanders, "Our Island, Being a Study of Pitcairn Community in Transition" (master's thesis, University of Auckland, 1953).

22. Claydon "Report on Pitcairn Island," 80–81.

23. Letter from A. Wotherspoon to Harry Brattain, Pitcairn Island, January 27, 1956, TNA, CO 1036/114.

24. Ernest Schubert, *The Pitcairn Island Story* (Longmans, 1961); Schubert et al., "Pitcairn Island Today," in Ross and Moverley, *The Pitcairnese Language*, 26–36.

25. Claydon, "Report on Pitcairn Island," 8.

26. "The 'Bounty' Bible Returns to Pitcairn Island," *Sydney Morning Herald*, April 22, 1950, 2.

27. Ross and Moverley, *The Pitcairnese Language*, 7.

28. Letter from the high commissioner to the secretary of state for the colonies, September 16, 1953, TNA, CO 1023/194.

29. Letter from G. Allen to the governor, Fiji, July 4, 1953, TNA, CO 1023/194.

30. E. G. Stanley, "Ross, Alan Strode Campbell (1907–1980)," in *Oxford Dictionary of National Biography*, ed. Lawrence Goldman, published September 23, 2004; last modified May 25, 2006, http://www.oxforddnb.com/view/article/69690.

31. Alan S. C. Ross, "Linguistic Class-Indicators in Present-Day English," *Neuphilologische Mitteilungen* 55, no. 1 (1954): 20–56; republished in Nancy Mitford, ed., *Noblesse Oblige: An Enquiry into the Identifiable Characteristics of the English Aristocracy* (Hamish Hamilton, 1956).

32. Jane Moverley wrote an account of her time on the island, published later as *Our Pitcairn Story* (pub. by author, 2007).

33. Letter from Jane Moverley to her MP, G. W. Lloyd, May 28, 1953, TNA, CO 1023/194.

34. Note by R. H. Holden, June 10, 1953, TNA, CO 1023/194.

35. Letter from R. Garvey, December 9, 1952, TNA, CO 1023/194.

36. Note to C. A. Kirkman, January 15, 1953, TNA, CO 1023/194.

37. Letter from A. W. Moverley to the Colonial Office, August 23, 1953, TNA, CO 1023/194.

38. A. S. C. Ross and A. W. Moverley, "Note on Some Pristine Place-Names of Pitcairn Island," in *Cinquième congrès international de toponymie et d'anthroponymie, Salamanca, 12–15 avril 1955: Actes et mémoires* [Fifth international congress of toponymy and anthroponymy,

Salamanca, April 12–15, 1955: Proceedings and transactions], vol. 2, ed. L. Cortés et al. (University of Salamanca, 1955).

39. Peter Edwards, "Some Reflections on Elwyn by Professor Peter Edwards," Flint Collection Finding Aid, Fryer Library, University of Queensland (July 1997), https://www.library.uq.edu.au/fryer/ms/Flint/flint_cat_edwards.html, accessed October 4, 2014.

40. Letter from Elwyn Flint to A. S. C. Ross, January 30, 1956, L.Add 1552.

41. I.e., letter from the Norfolk Island executive council to C. R. Pinney, July 18, 1932; and petition to the governor general of Australia, July 31, 1934, NLNZ, Micro-MS-Coll-08–0011.

42. O'Collins, *An Uneasy Relationship;* for a broader history by a local author, see Merval Hoare, *Norfolk Island: An Outline of Its History 1774–1987,* 4th ed. (University of Queensland Press, 1988).

43. C. R. Pinney to Kenneth Binns, December 23, 1937, Papers of Charles Robert Pinney, NLA, MS 9931.

44. Inspector Reay, "Education on Norfolk Island," May 30, 1912, 12, Flint Papers 4, folder 12.

45. Donald Stanley Marshall, "First and Second Expeditions," 1951, NLNZ, Micro-MS-Coll-08–1335–1. For a discussion of negative appraisals by outsiders, especially administrators and social theorists, see chapter 3.

46. E. H. Flint, "The Language of Norfolk Island," in Ross and Moverley, *The Pitcairnese Language,* 189.

47. "Dialogue 8," Transcriptions of Norfolk Island Dialogues, Flint Papers, box 3, folder 4.

48. Elwyn Flint, "Bilingual Interaction between Norfolk Island Language and English," paper presented at the First Conference of the Linguistic Circle of Canberra, May 20, 1961, in Papers of Elwyn Flint, UQFL 173, box 3, folder 1, Fryer Library, University of Queensland, St. Lucia (hereafter Flint Papers).

49. Flint, "Bilingual Interaction between Norfolk Island Language and English." Flint also described his preferred fieldwork methodology to New Zealand sociologist John Harré some years later; see letter from Flint to Harré, December 18, 1964, Flint Papers, box 4, folder 13.

50. Transcriptions of Norfolk Island Dialogues, Flint Papers, box 3, folder 4. For the IPA at the time and its development, see Robert W. Albright, *The International Phonetic Alphabet: Its Backgrounds and Development* (Indiana University Press, 1958); for a history of the International Phonetic Association, see M. K. C. Macmahon, "The International Phonetic Association: The First 100 Years," *Journal of the International Phonetic Association* 16, no. 1 (June 1986): 30–38.

51. E. H. Flint, "Norfolk Form-Meaning Reference List (FRL)," Flint Papers, box 4, folder 1.

52. Flint, "Bilingual Interaction between Norfolk Island Language and English," 4; the method was adopted from S. A. Wurm and D. C. Laycock, "The Question of Language and Dialect in New Guinea," *Oceania* 32, no. 2 (December 1961): 128–143.

53. Wurm and Laycock, "The Question of Language and Dialect in New Guinea," 133.

54. Flint, "Bilingual Interaction between Norfolk Island Language and English," 9.

55. Ross and Moverley, *The Pitcairnese Language;* see also reviews of it: J. E. Busein, review of *The Pitcairnese Language,* by A. S. C. Ross and A. W. Moverley, *Bulletin of the School of Oriental and African Studies* 28 (1965): 667–668; George W. Turner, review of *The Pitcairnese Language,* by A. S. C. Ross and A. W. Moverley, *Anglia* 86 (1968): 358–360; and Robert A. Hall, review of *The Pitcairnese Language,* by A. S. C. Ross and A. W. Moverley, *Journal of the Polynesian Society* 74 (1965): 137–138.

56. Ross and Moverley, *The Pitcairnese Language,* 11.

57. Turner, review of *The Pitcairnese Language,* 358.

58. Flint, "The Language of Norfolk Island," 189–211.

59. Flint, "The Language of Norfolk Island," 189.

60. Letter from Flint to Peter Gathercole, December 22, 1964, Flint Papers, box 4, folder 13.

61. Letter from Flint to Harré, December, 18, 1964, Flint Papers, box 4, folder 13; letter from Harré to Flint, October 14, 1965, Flint Papers, box 4, folder 13 (emphasis in original).

62. Letter from D. J. Rogers to Shirley Harrison, February 21, 1980, private archive of Peter Mühlhäusler.

63. M. Hoare, *Norfolk Island*, 23–24.

64. See M. Hoare, *Norfolk Island*, 23–24.

65. Shirley Harrison, "The Language of Norfolk Island" (master's thesis, Macquarie University, 1972).

66. Letter from Flint to John Harré, October 16, 1979, Flint Papers, box 4, folder 13.

67. Elwyn Henry Flint, "Stable Societal Diglossia in Norfolk Island," in *Sociolinguistic Studies in Language Contact: Methods and Cases*, ed. William Mackey and Jacob Ornstein (Mouton, 1979), 295–332.

68. Anne Harrison, "Letter to the *Norfolk Islander*," n.d., private archive of Anne Harrison; Shirley Harrison, "Variation in Present Norfolk Island Speech: A Study of Stability and Instability in Diglossia" (PhD diss., Macquarie University, 1984), 1.

69. Maude authored a chapter, "History of Pitcairn," in Ross and Moverley, *The Pitcairnese Language*, 45–101.

70. Letter from Henry Evans Maude to J. D. Davidson, September 14, 1960, Maude Papers, Special Collections, University of Adelaide (hereafter Maude Papers), box J, folder 1960 part II.

71. Obituary of Moresby Buffett, untitled, n.d., private archive of Peter Mühlhäusler.

72. Harrison, "Variation in Present Norfolk Island Speech," 184.

73. Shirley Harrison, "Points to Be Observed in Conduct of Interview," untitled notebook, n.d., presumably 1982, private archive of Peter Mühlhäusler; Harrison, "Varieties in Present Norfolk Island Speech," 92–93.

74. "Interview between Dad, Grace, John, Emily," untitled notebook, January 1982, private archive of Peter Mühlhäusler.

75. Shirley Harrison, "Topics Suited to N, Restriction of N," manila dated June 3, 1984, Harrison Papers, Norfolk Island Museum, 4 red.

76. Harrison, "Points to Be Observed in Conduct of Interview."

77. Harrison, "Variation in Present Norfolk Island Speech," 90–92.

78. Shirley Harrison, entry beginning "To this point," untitled notebook, n.d., presumably 1980, private archive of Peter Mühlhäusler.

79. Harrison, "November 27, 1980," *1980 Norfolk Notebook*, Harrison Papers, Norfolk Island Museum.

80. Letter from Harrison to Charlie Adams, June 13, 1980, private archive of Peter Mühlhäusler.

81. She translated fifteen sentences for a linguist at the University of Texas at Austin in 1982, for instance. Letter from Ian Hancock to Harrison, October 19, 1982, Harrison Papers, Norfolk Island Museum, 2 green.

82. Letter from Harrison to the principal, Norfolk Island public school, September 16, 1981, private archive of Peter Mühlhäusler.

83. Harrison, "November 25, 1980," *1980 Norfolk Notebook.*

84. Shirley Harrison, "August 31, 1970," untitled diary named in archive as "woolstar green exercise book" (hereafter *Norfolk Diary*), Harrison Papers, Norfolk Island Museum, 2 red.

85. Letter from Harrison to Joe Nobbs, n.d., private archive of Peter Mühlhäusler.

86. Harrison, "Variation in Present Norfolk Island Speech," 216.

87. Harrison, "Variation in Present Norfolk Island Speech," 81–82 (emphasis in original).

88. Harrison, "Variation in Present Norfolk Island Speech," 24.

89. Harrison, "Variation in Present Norfolk Island Speech," 8.

90. Harrison, "Variation in Present Norfolk Island Speech," 436. Similar arguments appear in subsequent papers and presentations: Shirley Harrison, "A Language History of Norfolk Island," in *Working Papers of the Speech and Language Research Centre 1.4* (School of English and Linguistics, Macquarie University, 1976); Shirley Harrison, "Bounty from the Mutiny," *Macquarie Dictionary Society* 4, no. 1 (1987): 2–4.

91. Shirley Harrison, "The Social Setting of Norfolk Speech," *English World-Wide* 6 (1985): 131–153.

92. Reinecke, "Pitcairnese and Norfolk"; letter from Shirley Harrison to John Reinecke, June 23, 1972, private archive of Peter Mühlhäusler; letter from Harrison to Reinecke, July 12, 1973, private archive of Joshua Nash; letter from Reinecke to Harrison, June 9, 1972, private archive of Joshua Nash; letter from Reinecke to Harrison, July 18, 1972, private archive of Joshua Nash.

93. John Holm, "Substrate Diffusion," in *Substrata versus Universals in Creole Genesis*, ed. Pieter Muysken and Norval Smith (John Benjamins, 1986), 266; John Holm, *Pidgins and Creoles*, vol. 2, *Reference Survey* (Cambridge University Press, 1989), 546–551.

94. Letter from Holm to Harrison, June 26, 1982, Harrison Papers, private archive of Joshua Nash.

95. Letter from Ian Hancock to Harrison, October 19 1982, Harrison Papers, Norfolk Island Museum, 2 green.

96. Letter from Manfred Gorlach to Harrison, April 27, 1979, private archive of Joshua Nash.

97. Arne Zettersten, "The Mutineers of the *Bounty* and the Language of Their Descendants," *Moderna Sprak* 60 (1966): 393–398.

98. He produced several publications: Anders Källgård, *Pitcairnese: A Report 30 Years after Moverley, Concentrating on the Vocabulary* (University of Gothenburg, 1981); Anders Källgård, *Fut yoli noo bin laane aklen? A Pitcairn Word List* [Why didn't you teach us? A Pitcairn word list] (University of Gothenburg, 1991); Anders Källgård, "A Pitkern World List," *Pacific Linguistics, Series A. Occasional Papers* 91 (1998): 107–171.

99. Letter from M. Gorlach to A. Zettersten, September 12, 1979, private archive of Joshua Nash.

100. Letter from Harrison to Zettersten, April 9, 1981, private archive of Joshua Nash.

101. Letter from Laycock to Harrison, April 15, 1986, Harrison Papers, Norfolk Island Museum, 15 red sleeve.

102. Letter from Laycock to Harrison, April 15, 1986, Harrison Papers, Norfolk Island Museum, 15 red sleeve.

103. Letter from Laycock to Harrison, April 28, 1987, Harrison Papers, Norfolk Island Museum, B orange.

104. Beryl Nobbs Palmer, *A Dictionary of Norfolk Words and Uses* (pub. by author, 1986).

105. Letter from Harrison to Laycock, n.d., probably a continuation of her letter dated May 14, 1987. The Harrison Papers contain draft copies of letters; I have preserved her edits and elisions.

106. Letter from Laycock to Harrison, April 28, 1987, Harrison Papers, Norfolk Island Museum, B orange.

107. Donald C. Laycock, "The Status of Pitcairn-Norfolk: Creole, Dialect, or Cant?" in *Status and Function of Languages and Language Varieties*, ed. Ulrich Ammon (Walter de Gruyter, 1989), 608.

108. Laycock, "The Status of Pitcairn-Norfolk," 617.

109. Laycock, "The Status of Pitcairn-Norfolk."

110. Letter from Harrison to Laycock, May 14, 1987, Harrison Papers, Norfolk Island Museum, B orange.

111. Letter from Harrison to Laycock, n.d., likely 1987, Norfolk Island Museum, 15 red sleeve.

112. Letter from Harrison to Laycock, n.d., likely 1987.

113. Lois Carrington, "Donald Clarence Laycock 1936–1988," *English World-Wide* 10, no. 1 (1989): 121–122.

114. Donald Laycock, "The Interpretation of Variation in Pitcairn-Norfolk," in *Development and Diversity: Language Variation across Time and Space: A Festschrift for Charles-James N. Bailey* (Summer Institute of Linguistics, 1990), 621–628.

115. Interview by the author with Professor Peter Mühlhäusler, Adelaide, March 30, 2014.

116. Peter Mühlhäusler, *Pitkern-Norf'k,* Dialects of English 17 (De Gruyter Mouton, 2020); Peter Mühlhäusler et al., *Ucklun's Norf'k: Words as a Memory of Our Past* (Norfolk Island Museum Trust, 2012).

117. Joshua Nash, *Insular Toponymies: Pristine Place-Naming on Norfolk Island, South Pacific and Dudley Peninsula, Kangaroo Island* (John Benjamins, 2013); Joshua Nash and Mitchell Low, "Language and Place-Knowledge on Norfolk Island," *Ethnos* 80, no. 3 (2015): 385–408; Joshua Nash, "Inside(r)-Outside(r): Toward a Linguistics and Sociology of Space on Pitcairn Island," *Australian Folklore* 31 (December 2, 2016), http://journals.kvasirpublishing.com/af/article/view/372; Joshua Nash, "Island Placenaming and Insular Toponymies," *Names* 63, no. 3 (September 1, 2015): 146–157, https://doi.org/10.1179/0027773815Z.000000000110.

118. Peter Mühlhäusler et al., *Ucklun's Norf'k.*

119. Email to the author from Piria Coleman, May 4, 2014.

120. "Norfolk Island Language (Norf'k) Bill," *Norfolk Island Hansard, 11th Assembly* (2004), 4.

121. Peter Mühlhäusler, "A Language Plan for Norfolk Island," in *Language Endangerment and Language Maintenance,* ed. David Bradley and Maya Bradley (Routledge Curzon, 2002); and Peter Mühlhäusler, "From Despised Jargon to Language of Education: Recent Developments in the Teaching of Norf'k," in *Education in Languages of Lesser Power: Asian-Pacific Perspectives,* ed. Craig Alan Volker and Fred. E. Anderson (John Benjamins, 2015), 223–241.

122. Cf. Andrei A. Avram, "Pitkern and Norfolk Revisited," *English Today* 19, no. 3 (July 2003): 44–49; Peter Mühlhäusler, "Some Notes on the Ontology of Norf'k," *Language Sciences* 33, no. 4 (July 2011): 673–679.

123. John Ingram and Peter Mühlhäusler, "Norfolk Island-Pitcairn English: Phonetics and Phonology," in *The Pacific and Australasia,* ed. Kate Burridge and Bernd Kortmann (Walter de Gruyter, 2008), 269.

124. Letter from Gorlach to Harrison, April 27, 1979, private archive of Peter Mühlhäusler.

125. Letter from Flint to J. Leighton, linguistic survey of Scotland, March 18, 1965, Flint Papers, box 3, folder 3.

126. Letter from Flint to R. Gardner White, July 24, 1961, Flint Papers, box 21, folder 3A.

127. Letter from Harrison to Flint, n.d., probably 1979, private archive of Peter Mühlhäusler.

128. Letter from A. R. Doherty to Harrison, February 15, 1984, private archive of Peter Mühlhäusler.

129. Letter to Peter Gathercole from Harrison, December 14, 1980, Harrison Papers, Norfolk Island Museum, 2 green.

130. A condition that recalls other collections produced by fieldwork; see Emma Kowal, "Orphan DNA: Indigenous Samples, Ethical Biovalue and Postcolonial Science," *Social Studies of Science* 43, no. 4 (August 2013): 577–597; Warwick Anderson, "The Frozen Archive, or Defrosting Derrida," *Journal of Cultural Economy* 8, no. 3 (May 4, 2015): 379–387, https://doi

.org/10.1080/17530350.2014.890637; Joanna Radin, *Life on Ice: A History of New Uses for Cold Blood* (University of Chicago Press, 2017).

131. Grimson, "The Pitcairnese Text," 124.

Conclusion

1. Miles C. Benton et al., "'Mutiny on the *Bounty*': The Genetic History of Norfolk Island Reveals Extreme Gender-Biased Admixture," *Investigative Genetics* 6, no. 11 (2015).

2. Justin C. Sherwin et al., "Prevalence of Chronic Ocular Diseases in a Genetic Isolate: The Norfolk Island Eye Study (NIES)," *Ophthalmic Epidemiology* 18, no. 2 (2011): 61–71, https://www.tandfonline.com/toc/iope20/18/2; C. Bellis, H. C. Cox, et al., "Linkage Mapping of CVD Risk Traits in the Isolated Norfolk Island Population," *Human Genetics* 124 (2008): 543–552, https://doi.org/10.1007/s00439–008–0580-y; Hannah C. Cox et al., "Heritability and Genome-Wide Linkage Analysis of Migraine in the Genetic Isolate of Norfolk Island," *Gene* 11, no. 1 (2011): 119–123.

3. McEvoy et al., "European and Polynesian Admixture in the Norfolk Island Population"; Stuart Macgregor et al., "Legacy of Mutiny on the *Bounty:* Founder Effect and Admixture on Norfolk Island," *European Journal of Human Genetics* 18 (2010): 67–72.

4. On links between genomic research and race science, see Jenny Reardon and Kim TallBear, "'Your DNA Is Our History': Genomics, Anthropology, and the Construction of Whiteness as Property," *Current Anthropology* 53, no. S5 (April 1, 2012): S233–S245, https://doi.org/10.1086/662629.

5. Dawn Oliver, "Problems on Pitcairn," in *Justice, Legality and the Rule of Law: Lessons from the Pitcairn Prosecutions*, ed. Dawn Oliver (Oxford University Press, 2009), 1–22.

6. Jan Corbett and Tony Stickley, "End of a Legend as Pitcairn Island Meets the Modern Law," *New Zealand Herald*, June 19, 2001, https://www.nzherald.co.nz/world/end-of-a-legend-as-pitcairn-island-meets-the-modern-law/FEIV2P4ROUOM3IJBHXWGCZDIZU/.

7. See also Lisa Fletcher, "Reading the News: Pitcairn Island at the Beginning of the 21st Century," *Island Studies Journal* 3, no. 1 (May 2008): 57–72.

8. Letter from R. J. Stratton to Sir Harold Smedley, October 8, 1979, TNA, FCO/107/114.

9. FCO (White) to PS/foreign secretary (Robin Cook) and PS/parliamentary under secretary (Baroness Patricia Scotland), May 8, 2000, in "Privy Council Record," eVols, University of Hawai'i—Manoa, https://evols.library.manoa.hawaii.edu/handle/10524/19431 (hereafter Privy Council Record). This collection was made available by Michael O. Eshleman. Pages in that record are noted first by volume, followed by a dash, then a page number, as in "8–3791."

10. Commissioner (Leon Salt) to governor (Martin Williams), October 24, 2000, Privy Council Record, 8–3779; commissioner (Salt) to acting-governor (Karen Wolstenholme), May 1, 2000, Privy Council Record, 8–3634.

11. PS (David Cairns)/parliamentary under secretary (Scotland) to PS/foreign secretary (Cook), May 10, 2000, Privy Council Record, 8–3646.

12. Presentation by mayor (Steve Christian) to Pacific Regional Seminar (Exhibit "E" to Gilbert Affidavit), May 14, 2002, Privy Council Record, 8–3937.

13. Statement by mayor's representative (K. Young) to Caribbean Regional Seminar (Exhibit "F" to Gilbert Affidavit), May 20, 2003, Privy Council Record, 8–3993.

14. United Nations Working Paper on Pitcairn (Exhibit "D" to Gilbert Affidavit), May 19, 2003, Privy Council Record, 8–3990.

15. Queen v 7 Named Accused [2004] Supreme Court of Pitcairn Islands (PNSC) 1, Pacific Islands Legal Information Institute, http://www.paclii.org/pn/cases/PNSC/2004/1.html, ¶11.

16. Queen v 7 Named Accused, ¶125.

17. Queen v 7 Named Accused, ¶59; judgment by Lord Hoffman, 10 (Christian and Ors v the Queen [2006] PNPC 1; [2006] United Kingdom Privy Council (UKPC) 47.

18. Judgment by Lord Hope, Christian and Ors v the Queen.

19. Judgment by Lord Hope, Christian and Ors v the Queen, ¶46, ¶48.

20. Oliver, *Justice, Legality, and the Rule of Law;* Michael O. Eshleman, "Law in Isolation: The Legal History of Pitcairn Island, 1900–2010," *ILSA Journal of International and Comparative Law* 18, no. 1 (2011): 17–72; Michael O. Eshleman, "A Preliminary Legal Bibliography of the Pitcairn Islands, South Pacific Ocean," *Law Library Journal* 106, no. 2 (2014): 221–236; Michael O. Eshleman, "A South Seas State of Nature: The Legal History of Pitcairn Island, 1790–1900," *Pacific Basin Law Journal* 29, no. 1 (January 2011), https://doi.org/10.5070/P8291022231; Helen Power, "Pitcairn Island: Sexual Offending, Cultural Difference, and Ignorance of the Law," *Criminal Law Review* (July 2007): 609; Anthony Trenwith, "The Empire Strikes Back: Human Rights and the Pitcairn Proceedings," *Journal of South Pacific Law* 7, no. 2 (2003).

21. Marks, *Lost Paradise;* see also William Prochnau and Laura Parker, "Trouble in Paradise," *Vanity Fair,* December 17, 2007, https://www.vanityfair.com/news/2008/01/pitcairn200801.

22. Marks, *Lost Paradise,* 166.

23. Marks, *Lost Paradise,* 272.

24. Marks, *Lost Paradise,* xxiii.

25. Shapiro, *The Heritage of the Bounty.*

26. H. E. Maude, "A General Report Based on Seven Months' Residence in the Island during 1940–41," 6, Maude Papers, part I, series J, section 14, 1941; "Illegitimates of Pitcairn Island," Maude Papers, part I, series A, Miscellaneous Correspondence and Papers.

27. Stoler, *Along the Archival Grain,* 24.

28. Rhiannon Adam, "Three Months on Pitcairn," *Journey of a Lifetime,* BBC Radio 4, October 19, 2015, https://www.bbc.co.uk/programmes/b06j144b; Adam, *Big Fence.*

29. Rhiannon Adam, "The Departure," *From London to Pitcairn* (blog), October 19, 2015, https://rhiannonsetsoff.wordpress.com/.

30. Adam, "The Departure."

31. Letter from Lord Chelmsford, July 20, 1910, 2, TNA, CO 537/463.

32. Marks, *Lost Paradise,* 16.

33. Dening, *Beach Crossings;* Dening, *Islands and Beaches.*

34. Vanessa Smith, "Pitcairn's Guilty Stock: The Island as Breeding Ground," in *Islands in History and Representation,* ed. Rod Edmond and Vanessa Smith (Routledge, 2003), 122.

35. Greg Fry, "Framing the Islands: Knowledge and Power in Changing Australian Images of 'the South Pacific,'" *Contemporary Pacific* 9, no. 2 (1997): 307.

36. Teresia K. Teaiwa, "L(o)osing the Edge," *Contemporary Pacific* 13, no. 2 (2001): 343–357.

37. Robert Langdon, "'Dusky Damsels': Pitcairn Island's Neglected Matriarchs of the Bounty Saga," *Journal of Pacific History* 35, no. 1 (2000): 29–47, https://doi.org/10.1080/713682826.

38. See especially Pauline Reynolds's work: Reynolds, "Tapa Cloths and Beaters"; Pauline Reynolds, "Pitcairn Tapa: Discovering the Past through Tapa," in *TAPA: From Tree Bark to Cloth: An Ancient Art of Oceania from Southeast Asia to Eastern Polynesia,* ed. Michel Charleux (Somogy Art Publishers, 2017), 370–375.

39. For a call to escape these boundaries on Pitcairn, see an article I authored jointly with six other scholars of Pitcairn Island: A. Young et al., "When Margins Are Centres."

40. Tuck and Yang, "R-Words: Refusing Research."

41. For more on alternative, Pacific historicities, see Chris Ballard, "Oceanic Historicities," *Contemporary Pacific* 26, no. 1 (2014): 96–124, https://doi.org/10.1353/cp.2014.0009; see also Miranda Johnson, "Toward a Genealogy of the Researcher as Subject in Post/Decolonial Pacific Histories," *History and Theory* 59, no. 3 (2020): 421–429, https://doi.org/10.1111/hith.12170; Warwick Anderson, "Finding Decolonial Metaphors in Postcolonial Histories," *History and Theory* 59, no. 3 (2020): 430–438, https://doi.org/10.1111/hith.12171.

Bibliography

Archives, Libraries, and Special Collections

Alexander Turnbull Library, National Library of New Zealand, Wellington
British Library, London
C. Templeton Crocker Papers, Special Collections, California Academy of Sciences Library, San Francisco
Cadbury Research Library, University of Birmingham, Birmingham
Caird Library and Archive, National Maritime Museum, Greenwich
Collections, Otago Museum, Dunedin
Elwyn Flint Papers, Fryer Library, University of Queensland, St. Lucia
Harry L. Shapiro Papers and Central Archives, Special Collections Library, American Museum of Natural History, New York
Henry Evans Maude Papers, Archives and Special Collections, University of Adelaide, Adelaide
Hocken Library, University of Otago, Dunedin
Library and Archives of the Norfolk Island Museum, Kingston, Norfolk Island
Manuscript Archive, Royal Geographical Society, London
Mitchell Library, State Library of New South Wales, Sydney
National Archives, Kew
National Library of Australia, Canberra
Papers of Sir Arthur Keith, Archives, Royal College of Surgeons, London
Pitcairn Islands Study Center, Pacific Union College, Angwin, California
Robert Henry Codrington Papers, Commonwealth and African Studies, Bodleian Library, Oxford

Private Collections

Much of Pitcairn's history is encoded in texts held in private archives, often in digital form on personal hard drives. In writing this book, I have become indebted to Pitcairn Islanders, Norfolk Islanders, and fellow researchers who have generously shared their material. Without the bounty of documents their kindness afforded, those whose personal papers are not archived in academic repositories would not have had their due. Unlike public collections cataloged by archivists, these documents often lack call numbers or rigid organizational systems. The papers of Shirley Harrison, on which much of chapter 5 is built, are a particularly special case. Her research on Norfolk Island, which spanned decades, left behind a treasure trove of linguistic and historical records. Some of these are in the possession of her daughter, others are in the temporary trust of linguists, while still others are held at the Norfolk Island Museum. I was able to consult them as photocopies and as digital files; I have attempted to mark each source as legibly as possible given the circumstances.

Hardwicke Knight Papers, private collection of Megan Davidson
Private archive of Meralda Warren
Private archive of Kevin Young
Shirley Harrison Papers, private collection of Anne Harrison
Shirley Harrison Papers, private collection of Peter Mühlhäusler
Shirley Harrison Papers, private collection of Joshua Nash

Published Sources

Adam, Rhiannon. *Big Fence.* Blow Up Press, 2022.

Addison, Catherine. "Gender and Race in Byron's *The Island* and Mitford's *Christina.*" *Antipodes* 18, no. 1 (June 2004): 72–76.

Albert, Donald Patrick. "Teehuteatuaonoa aka 'Jenny,' the Most Traveled Woman on the Bounty: Chronicling Female Agency and Island Movements with Google Earth." *Island Studies Journal* 16, no. 1 (May 2021): 190–208. https://doi.org/10.24043/isj.153.

Albright, Robert W. *The International Phonetic Alphabet: Its Backgrounds and Development.* Indiana University Press, 1958.

Alexander, Caroline. *The Bounty: The True Story of the Mutiny on the Bounty.* Viking, 2003.

Allina, Eric. "'Fallacious Mirrors': Colonial Anxiety and Images of African Labor in Mozambique, ca. 1929." *History in Africa* 24 (1997): 9–52. https://doi.org/10.2307/3172017.

Alpers, Edward A., and Matthew S. Hopper. "Speaking for Themselves? Understanding African Freed Slave Testimonies from the Western Indian Ocean, 1850s–1930s." *Journal of Indian Ocean World Studies* 1, no. 1 (September 29, 2017): 60–89. https://doi.org/10.26443/jiows.v1i1.20.

Altnik, Henrice. "An American Race Laboratory: Jamaica, 1865–1940." *Wadabagei: A Journal of the Caribbean and Its Diaspora* 10, no. 3 (Fall 2007): 53–83.

Amoamo, Maria. "Remoteness and Myth Making: Tourism Development on Pitcairn Island." *Tourism Planning and Development* 8, no. 1 (February 2011): 1–19. https://doi.org/10.1080/21568316.2011.554035.

Anderson, Atholl, and Peter White. "The Prehistoric Archaeology of Norfolk Island, Southwest Pacific." *Records of the Australian Museum, Supplement* 27 (November 28, 2001): 1–142. https://doi.org/10.3853/j.0812-7387.27.2001.1334.

Anderson, Warwick. *The Cultivation of Whiteness: Science, Health, and Racial Destiny in Australia.* Duke University Press, 2006.

Anderson, Warwick. "Finding Decolonial Metaphors in Postcolonial Histories." *History and Theory* 59, no. 3 (2020): 430–438. https://doi.org/10.1111/hith.12171.

Anderson, Warwick. "The Frozen Archive, or Defrosting Derrida." *Journal of Cultural Economy* 8, no. 3 (May 4, 2015): 379–387. https://doi.org/10.1080/17530350.2014.890637.

Anderson, Warwick. "Hybridity, Race, and Science: The Voyage of the *Zaca*, 1934–1935." *Isis* 103, no. 2 (June 1, 2012): 229–253.

Anderson, Warwick. "Marking Global Health History: The Postcolonial Worldliness of Biomedicine." *Social History of Medicine* 27, no. 2 (2014): 372–384.

Anderson, Warwick. "Racial Hybridity, Physical Anthropology, and Human Biology in the Colonial Laboratories of the United States." *Current Anthropology* 53, no. S5 (April 2012): S95–S107. https://doi.org/10.1086/662330.

Anderson, Warwick, and M. Susan Lindee. "Pacific Biologies: How Humans Become Genetic." *Historical Studies in the Natural Sciences* 50, no. 5 (November 2020): 483–497. https://doi.org/10.1525/hsns.2020.50.5.483.

Arneil, Barbara. *John Locke and America: The Defence of English Colonialism*. Clarendon Press, 1996.

Arnold, David. "'Illusory Riches': Representations of the Tropical World, 1840–1950." *Singapore Journal of Tropical Geography* 21, no. 1 (2000): 6–18.

Auerbach, Jeffrey A. *The Great Exhibition of 1851: A Nation on Display*. Yale University Press, 1999.

Avram, Andrei A. "Pitkern and Norfolk Revisited." *English Today* 19, no. 3 (July 2003): 44–49.

Ballard, Chris. "Oceanic Historicities." *Contemporary Pacific* 26, no. 1 (2014): 96–124. https://doi.org/10.1353/cp.2014.0009.

Bangham, Jenny, and Soraya de Chadarevian. "Human Heredity after 1945: Moving Populations Centre Stage." *Studies in History and Philosophy of Science Part C: Studies in History and Philosophy of Biological and Biomedical Sciences* 47 (September 1, 2014): 45–49. https://doi.org/10.1016/j.shpsc.2014.05.005.

Barkan, Elazar. *The Retreat of Scientific Racism: Changing Concepts of Race in Britain and the United States between the World Wars*. Cambridge University Press, 1992.

Barrow, John. *An Auto-biographical Memoir of Sir John Barrow, Bart., Late of the Admiralty: Including Reflections, Observations, and Reminiscences at Home and Abroad, from Early Life to Advanced Age*. John Murray, 1847.

Barrow, John. *A Description of Pitcairn's Island and Its Inhabitants: With an Authentic Account of the Mutiny of the Ship Bounty, and of the Subsequent Fortunes of the Mutineers*. J. and J. Harper, 1832.

Barrow, John. *The Eventful History of the Mutiny and Piratical Seizure of H.M.S. Bounty, Its Causes and Consequences*. J. Murray, 1831.

Barrow, John. *Interesting Narratives and Discoveries Including the Life of James Bruce, Esquire, Travels in South Africa, by J. Barrow, Esq., Discovery of an English Colony on Pitcairn's Island in the South Sea*. C. Hulbert, 1817.

Barrow, John. "Spirit of Discovery." *Mirror* 23, no. 647 (February 8, 1834): 83–86.

Barrow, John, A. A. Sandilands, and Joshua Hill. "Recent Accounts of the Pitcairn Islanders." *Journal of the Royal Geographical Society of London* 3 (January 1833): 156–168. https://doi.org/10.2307/1797603.

Bashford, Alison. *Imperial Hygiene: A Critical History of Colonialism, Nationalism and Public Health*. Palgrave Macmillan, 2003.

Beals, Alan Robin, George Dearborn Spindler, and Louise Schaubel Spindler. *Culture in Process*. Holt, Rinehart and Winston, 1967.

Beechey, Frederick William. *Narrative of a Voyage to the Pacific and Beering's Strait, to Co-operate with the Polar Expeditions, Performed in His Majesty's Ship "Blossom," under the Command of Captain F. W Beechey in the Years 1825, 26, 27, 28.* Henry Colburn and Richard Bentley, 1831.

Beer, Gillian. *Missionary Writing and Empire, 1800–1860.* Cambridge University Press, 2003.

Belcher, Diana Jolliffe. *The Mutineers of the Bounty and Their Descendants in Pitcairn and Norfolk Islands.* Harper and Brothers, 1871.

Bell, Duncan. *The Idea of Greater Britain: Empire and the Future of World Order, 1860–1900.* Princeton University Press, 2009.

Bell, Duncan. *Victorian Visions of Global Order: Empire and Relations in Nineteenth-Century Political Thought.* Cambridge University Press, 2007.

Bellis, C., H. Cox, T. Dyer, J. Charlesworth, K. Begley, S. Quinlan, R. Lea, S. Heath, J. Blangero, and L. Griffiths. "Linkage Mapping of CVD Risk Traits in the Isolated Norfolk Island Population." *Human Genetics* 124, no. 5 (2008): 543–52. https://doi.org/10.1007/s00439-008-0580-y.

Bellis, Claire, Roger M. Hughes, Kimberly N. Begley, Sharon Quinlan, Rod A. Lea, Simon C. Heath, John Blangero, and Lyn R. Griffiths. "Phenotypical Characterisation of the Isolated Norfolk Island Population Focusing on Epidemiological Indicators of Cardiovascular Disease." *Human Heredity* 60, no. 4 (2005): 211–219. https://doi.org/10.1159/000090545.

Benedict, Ruth. *Race: Science and Politics.* Viking, 1945.

Bennett, J. M. *Reluctant Democrat: Sir William Denison in Australia 1847–1861.* Federation Press, 2011.

Bennett, Tony. "Museum, Field, Colony." *Journal of Cultural Economy* 2, no. 1–2 (July 1, 2009): 99–116. https://doi.org/10.1080/17530350903064097.

Benton, Miles C., Shani Stuart, Claire Bellis, Donia Macartney-Coxson, David Eccles, Joanne E. Curran, Geoff Chambers, John Blangero, Rod A. Lea, and Lyn R. Griffiths. "'Mutiny on the Bounty': The Genetic History of Norfolk Island Reveals Extreme Gender-Biased Admixture." *Investigative Genetics* 6 (2015): 11.

Beshero-Bondar, Elisa E. "Romancing the Pacific Isles before Byron: Music, Sex, and Death in Mitford's *Christina*." *ELH* 76, no. 2 (2009): 277–308. https://doi.org/10.1353/elh.0.0049.

Bhabha, Homi. *The Location of Culture.* Routledge, 1994.

Birkett, Dea. *Serpent in Paradise.* Anchor Books, 1997.

Boas, Franz. "Changes in the Bodily Form of Descendants of Immigrants." *American Anthropologist* 14, no. 3 (1912): 530–562.

Boas, Franz. "Individual, Family, Population, and Race." *Proceedings of the American Philosophical Society* 87, no. 2 (August 16, 1943): 161–164.

Boas, Franz. *Race, Language, and Culture.* University of Chicago Press, 1940.

Boas, Franz. "Race Problems in America Today." *Science* 29, no. 752 (May 28, 1909): 839–849.

Booth, Alison. "Revisiting the Homes and Haunts of Mary Russell Mitford." *Nineteenth-Century Contexts* 30, no. 1 (2008): 39–65.

"The 'Bounty' Bible Returns to Pitcairn Island." *Sydney Morning Herald,* April 22, 1950.

Bramwell, Lord. *Nationalisation of Land: A Review of Mr. Henry George's "Progress and Poverty."* Liberty and Property Defence League, 1882.

Brantlinger, Patrick. *Dark Vanishings: Discourse on the Extinction of Primitive Races, 1800–1930.* Cornell University Press, 2003.

Brattain, Michelle. "Race, Racism, and Antiracism: UNESCO and the Politics of Presenting Science to the Postwar Public." *American Historical Review* 112, no. 5 (December 1, 2007): 1386–1413. https://doi.org/10.1086/ahr.112.5.1386.

Brown, Michael. "Cold Steel, Weak Flesh: Mechanism, Masculinity and the Anxieties of Late Victorian Empire." *Cultural and Social History* 14, no. 2 (March 15, 2017): 155–181. https://doi.org/10.1080/14780038.2016.1269538.

Bruchac, Margaret M. *Savage Kin: Indigenous Informants and American Anthropologists.* University of Arizona Press, 2018.

Buffett, John. "A Narrative of 20 Years' Residence on Pitcairn's Island." *Friend* 4, no. 1 (1846): 2–3.

Burnet, Macfarlane. "Migration and Race Mixture from the Genetic Angle." *Eugenics Review* 51, no. 2 (July 1959): 93–97.

Busein, J. E. Review of *The Pitcairnese Language* by A. S. C. Ross and A. W. Moverley. *Bulletin of the School of Oriental and African Studies* 28 (1965): 667–668.

Byron, George Gordon. *The Island, or Christian and His Comrades.* John Hunt, 1823.

Campbell, Chloe. *Race and Empire: Eugenics in Colonial Kenya.* Oxford University Press, 2012.

Carrington, Lois. "Donald Clarence Laycock 1936–1988." *English World-Wide* 10, no. 1 (1989): 121–122.

Carteret, Philip. "Captain Carteret's Voyage." In *An Account of the Voyages Undertaken by the Order of His Present Majesty for Making Discoveries in the Southern Hemisphere and Successively Performed by Commodore Byron, Captain Carteret, Captain Wallis, and Captain Cook in the Dolphin, the Swallow, and the Endeavour,* edited by John Hawkesworth. Vol. 1. Strahan and Cadell, 1773.

Caspari, Rachel. "1918: Three Perspectives on Race and Human Variation." *American Journal of Physical Anthropology* 139, no. 1 (2009): 5–15. https://doi.org/10.1002/ajpa.20975.

Castle, William E. *Genetics and Eugenics: A Textbook for Students of Biology and a Reference Book for Plant and Animal Breeders.* Harvard University Press, 1921.

Chamberlain, J. Edward, and Sander Gilman, eds. *Degeneration: The Dark Side of Progress.* Columbia University Press, 1985.

Chauvel, Charles, dir. *In the Wake of the Bounty.* Expeditionary Films, 1933.

Cheyne, Thomas Kelly. *The Relations between Civilized and Uncivilized Races: A Price Essay Read in the Theatre, Oxford, June 8, 1864.* T. and G. Shrimpton, 1864.

Clifford, James, and George E. Marcus. *Writing Culture: The Poetics and Politics of Ethnography.* University of California Press, 1986.

Colley, Linda. *Britons: Forging the Nation 1707–1837.* Yale University Press, 1992.

Colley, Linda. *The Gun, the Ship and the Pen: Warfare, Constitutions and the Making of the Modern World.* Liveright, 2021.

Collopy, Peter. "Race Relationships: Collegiality and Demarcation in Physical Anthropology." *Journal of the History of the Behavioral Sciences* 51, no. 3 (2015): 237–260.

Comos, Juan. *Manual of Physical Anthropology.* Charles Thomas, 1960.

Conklin, Alice. *In the Museum of Man: Race, Anthropology, and Empire in France, 1850–1950.* Cornell University Press, 2013.

Conner, Stuart W. "Constitutionality of Miscegenetic Marriages." *Montana Law Review* 11 (1950): 53.

The Converted Mutineer and His Bible Class: Or, John Adams and the Children of the Mutineers. Massachusetts Sabbath School Society, 1855.

Coon, Carleton. *The Origin of Races.* Alfred A. Knopf, 1963.

Coon, Carleton. *The Races of Europe.* Macmillan, 1939.

Coon, Carleton, Stanley Garn, and Joseph Birdsell. *Races: A Study of the Problems of Race Formation in Man.* Charles Thomas, 1950.

Corbett, Jan, and Tony Stickley. "End of a Legend as Pitcairn Island Meets the Modern Law." *New Zealand Herald,* June 19, 2001. https://www.nzherald.co.nz/world/end-of-a-legend-as-pitcairn-island-meets-the-modern-law/FEIV2P4ROUOM3IJBHXWGCZDIZU/.

"Correspondence on the Subject of Removal of Inhabitants of Pitcairn's Island to Norfolk Island." *British Parliamentary Papers.* HMSO, 1857.

"Correspondence Relating to the Transfer of Norfolk Island to the Government of New South Wales." *British Parliamentary Papers.* HMSO, 1897.

Corri, Montague. *A New Romantick Operatick Ballet Spectacle, Founded on the Recent Discovery of a Numerous Colony, Formed by, and Descended from the Mutineers of the Bounty Frigate Called Pitcairn's Island.* London, 1816. http://nla.gov.au/nla.obj-33744670.

Cox, Hannah C., Rod A. Lea, and Claire Bellis. "Heritability and Genome-Wide Linkage Analysis of Migraine in the Genetic Isolate of Norfolk Island." *Gene* 11, no. 1 (2011): 119–123.

Crawfurd, John. "On the Supposed Infecundity of Human Hybrids or Crosses." *Transactions of the Ethnological Society of London* 3 (January 1865): 356–362. https://doi.org/10.2307/3014171.

Crocker, Templeton. *The Cruise of the Zaca.* Harper, 1933.

Crook, Tony. *Anthropological Knowledge, Secrecy and Bolivip, Papua New Guinea: Exchanging Skin.* Oxford University Press, 2007.

Crook, Tony. "Exchanging Skin: Making a Science of the Relation between Bolivip and Barth." *Social Analysis* 53, no. 2 (Summer 2009): 94–107.

"A Cruise among the South Sea Islands." *Leisure Hour; an Illustrated Magazine for Home Reading,* 1890, 348–350.

Cummins, Jerold D., and John L. Kane Jr. "Miscegenation, the Constitution, and Science." *Dicta* 38 (1961): 24.

Currey, C. H. "Sir William Thomas Denison (1804–1871)." In *Australian Dictionary of Biography.* National Centre of Biography, Australian National University. Published 2006. http://adb.anu.edu.au/biography/denison-sir-william-thomas-3394/text5145.

Cutmore, Jonathan. "The Early 'Quarterly Review': New Attributions of Authorship." *Victorian Periodicals Review* 28, no. 4 (December 1, 1995): 305–329.

Daniels, Mary S. "The Mutineers of the 'Bounty.'" *Methodist Magazine* 36, no. 6 (December 1892): 565.

Darnell, Regna. *And Along Came Boas: Continuity and Revolution in Americanist Anthropology.* John Benjamins Publishing, 1998.

Darnell, Regna. *Invisible Genealogies: A History of Americanist Anthropology.* University of Nebraska Press, 2001.

Darnell, Regna. "The Professionalization of American Anthropology: A Case Study in the Sociology of Knowledge." *Social Science Information* 10, no. 2 (April 1971): 83–103.

Darwin, Charles. *The Descent of Man and Sexual Selection in Relation to Sex.* John Murray, 1871.

Davenport, Charles B. "The Effects of Race Intermingling." *Proceedings of the American Philosophical Society* 56, no. 4 (1917): 364–368.

Davenport, Charles B., and Morris Steggerda. *Race Crossing in Jamaica.* Carnegie Institution, 1929.

David, Deirdre. *Rule Britannia: Women, Empire, and Victorian Writing.* Cornell University Press, 1995.

Dawn, Oliver. "Problems on Pitcairn." In *Justice, Legality and the Rule of Law: Lessons from the Pitcairn Prosecutions,* edited by Oliver Dawn, 1–22. Oxford University Press, 2009.

Daws, Gavan. "Texts and Contexts: A First-Person Note." *Journal of Pacific History* 41, no. 2 (September 2006): 250.

DeCamp, David. "The Field of Creole Language Studies." *Latin American Research Review* 3, no. 3 (1968): 25–46.

"Degeneracy of Pitcairn Islanders." *Journal of the American Medical Association* 32, no. 14 (April 8, 1899): 771.

DeGraff, Michel. "Against Creole Exceptionalism." *Language* 79, no. 2 (June 2003): 391–410.

Delano, Amasa. *Narrative of Voyages and Travels in the Northern and Southern Hemispheres: Comprising Three Voyages round the World; Together with a Voyage of Survey and Discovery, in the Pacific Ocean and Oriental Islands.* E. G. House, 1817.

Dening, Greg. *Beach Crossings: Voyaging across Times, Cultures and Self.* Melbourne University Press, 2004.

Dening, Greg. *Islands and Beaches: Discourse on a Silent Land: Marquesas, 1774–1880.* Dorsey Press, 1980.

Dening, Greg. *Mr Bligh's Bad Language: Passion, Power and Theatre on the Bounty.* Cambridge University Press, 1994.

Denison, William Thomas. *Varieties of Vice-Regal Life.* 2 vols. Longmans, Green, 1870.

Dent, Rosanna. "Whose Home Is the Field?" *Isis* 113, no. 1 (March 2022): 137–143. https://doi.org/10.1086/718150.

"Descendants of the *Bounty*'s Crew, on Pitcairn's Island." *Calcutta Journal* 4, no. 133 (July 1819).

Dewbury, Adam. "The American School and Scientific Racism in Early American Anthropology." *Histories of Anthropology Annual* 3, no. 1 (2007): 121–147.

Diamond, Jared. *Collapse: How Societies Choose to Fail or Succeed.* Penguin, 2011.

Diamond, Jared. "Colonization Cycles in Man and Beast." *World Archaeology* 8, no. 3 (February 1977): 249–261. https://doi.org/10.1080/00438243.1977.9979671.

Diamond, Jared. "Foreword: Introduction to the Exploration of the Pitcairn Islands." *Biological Journal of the Linnean Society* 56, no. 1–2 (1995): 1–5. https://doi.org/10.1111/j.1095-8312.1995.tb01073.x.

Diamond, Jared. "Human Settlements: Why Did the Polynesians Abandon Their Mystery Islands?" *Nature* 317, no. 6040 (October 31, 1985): 764.

Diamond, Jared. "Pitcairn before the Bounty." *Nature* 369, no. 6482 (June 23, 1994): 23–24.

Diamond, Jared, and James A. Robinson, eds. *Natural Experiments of History.* Harvard University Press, 2010.

Diderot, Denis. "Supplément au voyage de Bougainville" [Addendum to the journey of Bougainville]. In *Diderot: Political Writings,* edited by John Hope Mason and Robert Wokler, 31–76. Cambridge Texts in the History of Political Thought. Cambridge University Press, 1992.

Dilke, Charles Wentworth. *Greater Britain: A Record of Travel in English-Speaking Countries during 1866 and 1867.* Macmillan, 1869.

Donaldson, Roger, dir. *The Bounty.* Orion Pictures, 1984.

Doubleday, Thomas. *The True Law of Population Shewn to Be Connected with the Food of the People.* Simpkin, Marshall, 1842.

Douglas, Bronwen. *Science, Voyages, and Encounters in Oceania, 1511–1850.* Palgrave Macmillan, 2014.

Douglas, Bronwen, and Chris Ballard. "Contact Tracing: The Materiality of Encounters." *History and Anthropology* 33, no. 1 (January 1, 2022): 1–16. https://doi.org/10.1080/02757206.2021.1987234.

Douglas, Bronwen, and Chris Ballard, eds. *Foreign Bodies: Oceania and the Science of Race 1750–1940.* Australian National University Press, 2008.

Douglas, Bronwen, and Doug Munro. "Of Islands and Sandalwood: Shineberg, Maude, and the Hidden History of Trade." In *Texts and Contexts: Reflections in Pacific Islands Historiography,* edited by Brij V. Lal and Doug Munro, 140–153. University of Hawai'i Press, 2005.

Driver, Felix, and Luciana Martins. *Tropical Visions in an Age of Empire.* University of Chicago Press, 2005.

Dunn, Leslie C. "An Anthropometric Study of Hawaiians of Pure and Mixed Blood, Based upon Data Collected by Alfred M. Tozzer." *Papers of the Peabody Museum of American Archaeology and Ethnology, Harvard University* 11, no. 3 (1928): 89–211.

Dyke, Henry. "Letter from Henry Dyke to the Secretary of the Admiralty, November 23, 1897, Enclosure No. 1." In *Pitcairn Island: Correspondence Relating to the Condition of the Pitcairn Islanders.* HMSO, 1899.

Edmond, Rod. *Representing the South Pacific: Colonial Discourse from Cook to Gauguin.* Cambridge University Press, 1997.

Edwards, Julia. "Phosphate and Forced Relocation: An Assessment of the Resettlement of the Banabans to Northern Fiji in 1945." *Journal of Imperial and Commonwealth History* 41, no. 5 (December 1, 2013): 783–803. https://doi.org/10.1080/03086534.2013.814252.

Edwards, Julia. "Phosphate Mining and the Relocation of the Banabans to Northern Fiji in 1945: Lessons for Climate Change-Forced Displacement." *Journal de la Société des Océanistes,* no. 138–139 (December 15, 2014): 121–136. https://doi.org/10.4000/jso.7100.

Edwards, Peter. "Some Reflections on Elwyn by Professor Peter Edwards." Flint Collection Finding Aid, Fryer Library, University of Queensland. July 1997. Accessed October 4, 2014. https://www.library.uq.edu.au/fryer/ms/Flint/flint_cat_edwards.html.

"Ein Brief aus Pitcairn" [A letter from Pitcairn]. In *Evangelisches Missions-Magazin,* 202–213. Verlag der Missionsbuchhandlung, 1883.

Elleray, Michelle. "Crossing the Beach: A Victorian Tale Adrift in the Pacific." *Victorian Studies* 47, no. 2 (Winter 2005): 164–173.

Errington, Joseph. "Colonial Linguistics." *Annual Review of Anthropology* 30 (2001): 19–39.

Erskine, Nigel. "The Historical Archaeology of Settlement at Pitcairn Island 1790–1856." PhD diss., James Cook University, 2004. http://eprints.jcu.edu.au/1259/.

Erskine, Nigel. "The Pitcairn Project: A Preliminary Report of the First Integrated Archaeological Investigation of the Mutineer Settlement of Pitcairn Island." *Bulletin of the Australian Institute for Maritime Archaeology* 23 (1999): 3–9.

Eshleman, Michael O. "Law in Isolation: The Legal History of Pitcairn Island, 1900–2010." *ILSA Journal of International and Comparative Law* 18, no. 1 (2011): 17–71.

Eshleman, Michael O. "A Preliminary Legal Bibliography of the Pitcairn Islands, South Pacific Ocean." *Law Library Journal* 106, no. 2 (2014): 221–236.

Eshleman, Michael O. "A South Seas State of Nature: The Legal History of Pitcairn Island, 1790–1900." *UCLA Pacific Basin Law Journal* 29, no. 1 (2011). https://doi.org/10.5070/P8291022231.

Fabian, Johannes. *Out of Our Minds: Reason and Madness in the Exploration of Central Africa*. University of California Press, 2000.

Fabian, Johannes. *Time and the Other: How Anthropology Makes Its Object*. Columbia University Press, 2002.

Fangerau, Heiner. "Rassenhygiene und Öffentlichkeiten: Die Popularisierung des rassenhygienischen Werkes von Erwin Baur, Eugen Fischer und Fritz Lenz" [Racial hygiene and the public: The popularization of racial hygiene works by Erwin Baur, Eugen Fischer and Fritz Lenz]. In *Bevölkerungsfragen: Prozesse des Wissenschaftstransfers in Deutschland und Frankreich 1870–1939* [Population questions: Processes of knowledge transfers in Germany and France], edited by Patrick Krassnitzer and Petra Overath, 131–154. Bölau, 2007.

Farber, Paul. "Race-Mixing and Science in the United States." *Endeavour* 27, no. 4 (December 2003): 166–170.

Farrah, F. W. "On Hybridity." *Journal of the Anthropological Society of London* 2 (1864): 222–224.

Felber, Lynette, ed. *Clio's Daughters: British Women Making History, 1790–1899*. University of Delaware Press, 2007.

Fischer, Eugen. *Die Rehebother Bastards und das Bastardierungsproblem beim Menschen* [The Rehebother basters and the race mixture problem in humans]. Gustav Fischer, 1913.

Fiske, N. W. *Aleck: The Last of the Mutineers, or the History of Pitcairn's Island*. E. C. Biddle, 1845.

Fletcher, Lisa. "Reading the News: Pitcairn Island at the Beginning of the 21st Century." *Island Studies Journal* 3, no. 1 (May 2008): 57–72.

Flint, Elwyn H. "The Language of Norfolk Island." In Alan S. C. Ross and A. W. Moverley, *The Pitcairnese Language*, 189–211. Andre Deutsch, 1964.

Flint, Elwyn H. "Stable Societal Diglossia in Norfolk Island." In *Sociolinguistic Studies in Language Contact: Methods and Cases*, edited by William Mackey and Jacob Ornstein, 295–332. Mouton, 1979.

Ford, Herbert. *Island of Tears: John I. Tay and the Story of Pitcairn*. Pacific Press, 1989.

Ford, Herbert. *Pitcairn Island as a Port of Call: A Record, 1790–2010*. 2nd ed. McFarland, 2012.

Ford, Herbert. *Pitcairn: Port of Call*. Hawser Titles, 1996.

Foucault, Michel. *The Order of Things*. Vintage Books, 1971.

Fraiture, Pierre-Philippe. *Past Imperfect: Time and African Decolonization, 1945–1960*. Oxford University Press, 2021.

Fry, Greg. "Framing the Islands: Knowledge and Power in Changing Australian Images of 'the South Pacific.'" *Contemporary Pacific* 9, no. 2 (1997): 305–344.

"Further Correspondence on the Subject of Removal of Inhabitants of Pitcairn's Island to Norfolk Island." *British Parliamentary Papers*. HMSO, 1857.

Ganett, Lisa. "Racism and Human Genome Diversity Research: The Ethical Limits of Population Thinking." *Philosophy of Science* 68, no. 3 (September 2001): 479–487.

Garn, Stanley, and Carleton Coon. "On the Number of Races of Mankind." *American Anthropologist* 57, no. 5 (October 1955): 996–1001.

Garritzen, Elise. *Pasha and His Historic Harem: Edward A. Freeman, Edith Thompson and the Gendered Personae of Late-Victorian Historians.* Manchester University Press, 2019.

Garritzen, Elise. "Women Historians, Gender and Fashioning the Authoritative Self in Paratexts in Late-Victorian Britain." *Women's History Review* 30, no. 4 (June 7, 2021): 650–668.

George, Henry. *Progress and Poverty: An Inquiry into the Cause of Industrial Depressions and of Increase in Want with Increase in Wealth, the Remedy.* William Reeves, 1884.

Gessler, Bernhard. *Eugen Fischer, 1874–1967: Leben und Werk des Freiburger Anatomen, Anthropologen und Rassenhygienikers bis 1927* [Eugen Fischer: 1874–1967: The life and work of the Freiburg anatomist, anthropologist, and race hygenicist until 1927]. Lang, 2000.

Gibbs, Martin, Brad Duncan, and Robert Varman. "The Free and Unfree Settlements of Norfolk Island: An Overview of Archaeological Research." *Australian Archaeology* 83, no. 3 (September 2, 2017): 82–99. https://doi.org/10.1080/03122417.2017.1404732.

Gikandi, Simon. *Maps of Englishness: Writing Identity in the Culture of Colonialism.* Columbia University Press, 1996.

Giles, Eugene. "Two Faces of Earnest A. Hooton." *American Journal of Physical Anthropology* 149, no. S55 (January 1, 2012): 105–113.

Gillian, John Phillip. *The Ways of Men: An Introduction to Anthropology.* Appleton-Century-Crofts, 1948.

Gil-Riaño, Sebastián. "Relocating Anti-Racist Science: The 1950 UNESCO Statement on Race and Economic Development in the Global South." *British Journal for the History of Science* 51, no. 2 (June 2018): 281–303. https://doi.org/10.1017/S0007087418000286.

Gonschor, Lorenz. "Norfolk Island." *Contemporary Pacific* 29, no. 1 (2017): 154–165. https://doi.org/10.1353/cp.2017.0011.

Gossler, Claus. "The Commercial Career of James Lyle Young: Contemporary Evaluations of a Notable South Seas Character." *Journal of Pacific History* 44, no. 2 (September 2009): 209–220.

Gould, Stephen Jay. *The Mismeasure of Man.* W. W. Norton, 1996.

Graham, Sylvester. *A Lecture to Young Men on Chastity: Intended Also for the Serious Consideration of Parents and Guardians.* G. W. Light, 1838.

Greenslade, William. *Degeneration, Culture, and the Novel, 1880–1940.* Cambridge University Press, 1994.

Grimson, A. C. "The Pitcairnese Text." In Alan S. C. Ross and A. W. Moverley, *The Pitcairnese Language,* 121–135. Andre Deutsch, 1964.

Grove, Richard H. *Green Imperialism: Colonial Expansion, Tropical Island Edens and the Origins of Environmentalism, 1600–1860.* Cambridge University Press, 1996.

Guggenheim, Michael. "Laboratizing and De-Laboratizing the World: Changing Sociological Concepts for Places of Knowledge Production." *History of the Human Sciences* 25, no. 1 (February 1, 2012): 99–118. https://doi.org/10.1177/0952695111422978.

Gunson, Niel, ed. *The Changing Pacific: Essays in Honour of H. E. Maude.* Oxford University Press, 1978.

Gunson, Niel. "Harry Maude: Unimane, Statesman and Pacific Historian." *Journal of Pacific History* 42, no. 1 (2007): 109–118.

Gurwit, S. Gordon. "Out of the Coral Sea." *Argosy All-Story Weekly* 147, no. 2 (November 18, 1922): 186.

Hall, James Norman. *Tale of a Shipwreck.* Houghton Mifflin, 1934.

Hall, Robert A. Review of *The Pitcairnese Language,* by A. S. C. Ross and A. W. Moverley. *Journal of the Polynesian Society* 74, no. 1 (1965): 137–138.

Haraway, Donna. *Primate Visions: Gender, Race, and Nature in the World of Modern Science.* Routledge, 1990.

Harms, Erik, Shafqat Hussain, Sasha Newell, Charles Piot, Louisa Schein, Sara Shneiderman, Terence Turner, and Juan Zhang. "Remote and Edgy: New Takes on Old Anthropological Themes." *HAU: Journal of Ethnographic Theory* 4, no. 1 (June 2014): 361–381. https://doi.org/10.14318/hau4.1.020.

Harrison, Shirley. "Bounty from the Mutiny." *Macquarie Dictionary Society* 4, no. 1 (1987): 2–4.

Harrison, Shirley. "The Language of Norfolk Island." Master's thesis, Macquarie University, 1972.

Harrison, Shirley. "A Language History of Norfolk Island." *Speech and Language Research Centre Working Papers* 1, no. 4 (1976).

Harrison, Shirley. "A Language History of Norfolk Island." In *Working Papers of the Speech and Language Research Centre 1.4.* School of English and Linguistics, Macquarie University, 1976.

Harrison, Shirley. "The Social Setting of Norfolk Speech." *English World-Wide* 6 (1985): 131–153.

Harrison, Shirley. "Variation on Present Norfolk Island Speech: A Study of Stability and Instability in Diglossia." PhD diss., Macquarie University, 1984.

Hasse, Stina. "I Am Sitting in a Room." *Body, Space and Technology* 11 (January 1, 2012). https://doi.org/10.16995/bst.71.

Hau'ofa, Epeli. "Our Sea of Islands." *Contemporary Pacific* 6, no. 1 (1994): 148–161.

Hazard, Anthony Q. "A Racialized Deconstruction? Ashley Montagu and the 1950 UNESCO Statement on Race." *Transforming Anthropology* 19, no. 2 (October 1, 2011): 174–186.

Hennessy, Elizabeth. "The Politics of a Natural Laboratory: Claiming Territory and Governing Life in the Galápagos Islands." *Social Studies of Science* 48, no. 4 (August 2018): 483–506. https://doi.org/10.1177/0306312718788179.

Herskovits, Melville J. "A Critical Discussion of the 'Mulatto Hypothesis.'" *Journal of Negro Education* 3, no. 3 (July 1, 1934): 389–402.

Hesketh, Ian. *The Science of History in Victorian Britain: Making the Past Speak.* Science and Culture in the 19th Century 12. Routledge, 2015.

Heyerdahl, Thor. *Aku-Aku: The Secret of Easter Island.* Rand McNally, 1958.

Heyerdahl, Thor. "The Voyage of the Raft of Kon-Tiki." *Geographic Journal* 115, no. 1/3 (1950): 20–41.

Heyerdahl, Thor, and Edwin N. Ferdon. *Reports of the Norwegian Archaeological Expedition to Easter Island and the East Pacific.* Rand McNally, 1965.

Hickson, William Edward. *Malthus: An Essay on the Principle of Population in Refutation of the Theory of the Rev. T. R. Malthus.* Walton, and Maberly, 1849.

"A History of Pitcairn Island." *Signs of the Times* 20, no. 54 (November 19, 1894).

Hoare, Merval. *Norfolk Island: An Outline of Its History 1774–1987.* 4th ed. University of Queensland Press, 1988.

Hoare, Nicholas. "Re-mining Makatea: People, Politics and Phosphate Rock." PhD diss., Australian National University, 2020. https://www.proquest.com/docview/2564078845/abstract/F25C5F2FEE7E4EF3PQ/1.

Holm, John. *An Introduction to Pidgins and Creoles.* Cambridge University Press, 2000.

Holm, John. *Pidgins and Creoles.* Vol. 2, *Reference Survey.* Cambridge University Press, 1989.

Holm, John. "Substrate Diffusion." In *Substrata versus Universals in Creole Genesis,* edited by Pieter Muysken and Norval Smith, 259–278. John Benjamins, 1986.

Hooton, Earnest A. "Plain Statements about Race." *Science* 83, no. 2161 (May 29, 1936): 511–513.

Hooton, Earnest A. "Progress in the Study of Race Mixtures with Special Reference to Work Carried On at Harvard University." *Proceedings of the American Philosophical Society* 65 (1926): 312–325.

Hurley, Kelly. *The Gothic Body: Sexuality, Materialism, and Degeneration in the Fin-de-Siecle.* Cambridge University Press, 1996.

Hussain, Shafqat. *Remoteness and Modernity: Transformation and Continuity in Northern Pakistan.* Yale University Press, 2015.

Huzel, James P. *The Popularization of Malthus in Early Nineteenth-Century England: Martineau, Cobbett and the Pauper Press.* Ashgate, 2006.

Hymes, Dell H. *Essays in the History of Linguistic Anthropology.* John Benjamins Publishing, 1983.

Imada, Adria L. *An Archive of Skin, an Archive of Kin: Disability and Life-Making during Medical Incarceration.* University of California Press, 2022.

Ingram, John, and Peter Mühlhäusler. "Norfolk Island-Pitcairn English: Phonetics and Phonology." In *The Pacific and Australasia,* edited by Kate Burridge and Bernd Kortmann, 267–291. Walter de Gruyter, 2008.

Jablonski, Sebastian. "The 'Uncorrupted' Paradise: Religion and Imperial Epistemic Violence on Pitcairn Island." In "New Scholarship in New Zealand and Pacific Studies Part 2," special issue, *Journal of New Zealand and Pacific Studies* 9 (December 1, 2021): 199–214. https://doi.org/10.1386/nzps_00070_1.

Jackson, John P., Jr. "'In Ways Unacademical': The Reception of Carleton S. Coon's *The Origin of Races.*" *Journal of the History of Biology* 34, no. 2 (June 2001): 247–285.

Jackson, John P., Jr. *Science for Segregation: Race, Law, and the Case against Brown v. Board of Education.* New York University Press, 2005.

Jackson, John P., and Nadine M. Weidman. *Race, Racism, and Science: Social Impact and Interaction.* ABC-CLIO, 2004.

Johnson, Adam. "Secretsharers: Intersecting Systems of Knowledge and the Politics of Documentation in Southwesternist Anthropology, 1880–1930." PhD diss., University of Michigan, 2018. https://deepblue.lib.umich.edu/handle/2027.42/146056.

Johnson, Miranda. "Toward a Genealogy of the Researcher as Subject in Post/Decolonial Pacific Histories." *History and Theory* 59, no. 3 (2020): 421–429. https://doi.org/10.1111/hith.12170.

"Journal of a Cruise Made to the Pacific Ocean, by Captain David Porter." *Quarterly Review* 13, no. 26 (July 1815): 374–383.

Jurmain, Robert, Lynn Kilgore, Wenda Trevathan, and Russell L. Ciochon, eds. *Introduction to Physical Anthropology.* Wadsworth, 2011.

Källgård, Anders. *Fut Yoli Noo Bin Laane Aklen? A Pitcairn Word List* [Why didn't you teach us? A Pitcairn word list]. University of Gothenburg, 1991.

Källgård, Anders. *Pitcairnese: A Report 30 Years after Moverley, Concentrating on the Vocabulary.* University of Gothenburg, 1981.

Källgård, Anders. "A Pitkern World List." In *Pacific Linguistics, Series A. Occasional Papers* 91 (1988): 107–171.

Källgård, Anders. "Present-Day Pitcairnese." *English World-Wide* 14, no. 1 (July 10, 2012): 71–114.

Keel, Terence D. "Religion, Polygenism and the Early Science of Human Origins." *History of the Human Sciences* 26, no. 2 (April 2013): 3–32.

Keith, Arthur. "The Physical Characteristics of Two Pitcairn Islanders." *Man* 17, no. 88 (August 1917): 121–131. https://doi.org/10.2307/2788792.

Kenny, Robert. "From the Curse of Ham to the Curse of Nature: The Influence of Natural Selection on the Debate on Human Unity before the Publication of *The Descent of Man*." *British Journal for the History of Science* 40, no. 3 (September 2007): 367–388.

King, Charles. *Gods of the Upper Air: How a Circle of Renegade Anthropologists Reinvented Race, Sex, and Gender in the Twentieth Century*. Knopf Doubleday, 2020.

King, Rachel. "Living on Edge: New Perspectives on Anxiety, Refuge and Colonialism in Southern Africa." *Cambridge Archaeological Journal* 27, no. 3 (August 2017): 533–551. https://doi.org/10.1017/S0959774317000312.

Kirch, Patrick V. "Microcosmic Histories: Island Perspectives on 'Global' Change." *American Anthropologist* 99, no. 1 (1997): 30–42. https://doi.org/10.1525/aa.1997.99.1.30.

Kittrell, Edward R. "Wakefield's Scheme of Systematic Colonization and Classical Economics." *American Journal of Economics and Sociology* 32, no. 1 (1973): 87–111.

Koerner, E. F. K. *Essays in the History of Linguistics*. John Benjamins Publishing, 2004.

Kohler, Robert E. *Landscapes and Labscapes: Exploring the Lab-Field Border in Biology*. University of Chicago Press, 2002.

Konishi, Shino. "Idle Men: The Eighteenth-Century Roots of the Indigenous Indolence Myth." In *Passionate Histories: Myth, Memory, and Indigenous Australia*, edited by Frances Peters-Little, Ann Curthoys, and John Docker, 99–122. Aboriginal History Monographs 21. Australian National University Press, 2010.

Kowal, Emma. "Orphan DNA: Indigenous Samples, Ethical Biovalue and Postcolonial Science." *Social Studies of Science* 43, no. 4 (August 2013): 577–597.

Krueger, Christine L. "Why She Lived at the PRO: Mary Anne Everett Green and the Profession of History." *Journal of British Studies* 42, no. 1 (January 2003): 65–90. https://doi.org/10.1086/342686.

Kuklick, Henrika. "Islands in the Pacific: Darwinian Biogeography and British Anthropology." *American Ethnologist* 23, no. 3 (1996): 611–638. https://doi.org/10.1525/ae.1996.23.3.02a00090.

Kuklick, Henrika, and Robert E. Kohler. *Science in the Field*. Osiris 11. University of Chicago Press, 1996.

Kumar, Krishnan. *The Making of English National Identity*. Cambridge University Press, 2003.

Lamb, Jonathan. *Preserving the Self in the South Seas, 1680–1840*. University of Chicago Press, 2001.

Langdon, Robert. "'Dusky Damsels': Pitcairn Island's Neglected Matriarchs of the Bounty Saga." *Journal of Pacific History* 35, no. 1 (2000): 29–47. https://doi.org/10.1080/713682826.

Langdon, Robert. "Harry Maude: Shy Proconsul, Dedicated Pacific Historian." In *The Changing Pacific: Essays in Honour of H. E. Maude*, edited by Niel Gunson, 1–21. Oxford University Press, 1978.

Largeaud-Ortega, Sylvie, ed. *The Bounty from the Beach: Cross-Cultural and Cross-Disciplinary Essays.* Australian National University Press, 2018. https://doi.org/10.22459/BB.10.2018.

Largeaud-Ortega, Sylvie. "Nordhoff and Hall's Mutiny on the Bounty: A Piece of Colonial Historical Fiction." In *The Bounty from the Beach: Cross-Cultural and Cross-Disciplinary Essays,* edited by Sylvie Largeaud-Ortega. Australian National University Press, 2018. https://doi.org/10.22459/BB.10.2018.

Lasker, Gabriel. *The Evolution of Man: A Brief Introduction to Physical Anthropology.* Holt, 1961.

Latham, Tim. *Norfolk: Island of Secrets; The Mystery of Janelle Patton's Death.* Allen and Unwin, 2005.

Latour, Bruno. *Science in Action: How to Follow Scientists and Engineers through Society.* Repr. ed. Harvard University Press, 1988.

Lavachery, Henri. "Contribution à l'étude de l'archéologie de l'île de Pitcairn" [Contribution to the study of the archaeology of Pitcairn Island]. *Bulletin de la Société des Américanistes de Belgique* 19 (1936): 3–42.

Laycock, Donald. "The Interpretation of Variation in Pitcairn-Norfolk." In *Development and Diversity: Language Variation across Time and Space: A Festschrift for Charles-James N. Bailey,* 621–628. Summer Institute of Linguistics, 1990.

Laycock, Donald. "The Status of Pitcairn-Norfolk: Creole, Dialect, or Cant?" In *Status and Function of Languages and Language Varieties,* edited by Ulrich Ammon, 608–629. Walter de Gruyter, 1989.

"Lesson 33: Story of Pitcairn Island." *Youth's Instructor,* May 23, 1911.

Levine, Philippa. *Prostitution, Race and Politics: Policing Venereal Disease in the British Empire.* Routledge, 2003.

Lewin, Terence. *Mutiny on the Bounty, 1789–1989: An International Exhibition to Mark the 200th Anniversary, 28 April 1989–1 Oct. 1989.* Manorial Research PLC in association with the National Maritime Museum, 1989.

Lewis, Joanna. *Empire of Sentiment: The Death of Livingstone and the Myth of Victorian Imperialism.* Cambridge University Press, 2018.

Liebersohn, Harry. *The Travelers' World: Europe to the Pacific.* Harvard University Press, 2006.

Lindee, Susan, and Ricardo Ventura Santos. "The Biological Anthropology of Living Human Populations: World Histories, National Styles, and International Networks: An Introduction to Supplement 5." *Current Anthropology* 53, no. 5 (April 2012): S3–S16.

Little, Michael A. "Human Population Biology in the Second Half of the Twentieth Century." *Current Anthropology* 53, no. S5 (April 2012): S126–S138. https://doi.org/10.1086/662440.

Little, Michael A., and Kenneth A. R. Kennedy, eds. *Histories of American Physical Anthropology in the Twentieth Century.* Lexington, 2009.

Livingstone, Frank B. "Genetics, Ecology and the Origins of Incest and Exogamy." *Current Anthropology* 10, no. 1 (February 1969): 45–61.

Lloyd, Christopher. *Mr. Barrow of the Admiralty: A Life of Sir John Barrow, 1764–1848.* Collins, 1970.

Lloyd, Frank, dir. *Mutiny on the Bounty.* Metro-Goldwyn-Mayer (MGM), 1935.

Longford, Raymond, dir. *The Mutiny of the Bounty.* Crick and Jones, 1916.

Looser, Devoney. *British Women Writers and the Writing of History, 1670–1820.* Johns Hopkins University Press, 2005.

Lösch, Niels C. *Rasse als Konstrukt: Leben und Werk Eugen Fischers* [Race as construct: The life and work of Eugen Fischer]. Lang, 1997.

Lowe, Abraham Thompson. *Second Class Book: Principally Consisting of Historical, Geographical, and Biographical Lessons, Adapted to the Capacities of Youth, and Designed for Their Improvement: Intended for the Use of Schools.* Dorr and Howland, 1829.

Ludovici, Anthony M. "Eugenics and Consanguineous Marriages." *Eugenics Review* 25, no. 3 (October 1933): 147–155.

Lummis, Trevor. *Pitcairn Island: Life and Death in Eden.* Ashgate, 1997.

Lutkehaus, Nancy Margaret. *Mead: The Making of an American Icon.* Princeton University Press, 2008.

Macgregor, Stuart, Claire Bellis, Rod A. Lea, Hanna Cox, Tom Dyer, John Blangero, Peter M. Visscher, and Lyn R. Griffiths. "Legacy of Mutiny on the *Bounty:* Founder Effect and Admixture on Norfolk Island." *European Journal of Human Genetics* 18 (2010): 67–72.

Mackey, David A., Justin C. Sherwin, Lisa S. Kearns, Yaling Ma, John Kelly, Byoung-Sun Chu, Robert Macmillan, et al. "The Norfolk Island Eye Study (NIES): Rationale, Methodology and Distribution of Ocular Biometry (Biometry of the Bounty)." *Twin Research and Human Genetics: The Official Journal of the International Society for Twin Studies* 14, no. 1 (February 2011): 42–52. https://doi.org/10.1375/twin.14.1.42.

Macklin, Robert Dark. *Paradise: Norfolk Island; Isolation, Savagery, Mystery and Murder.* Hachette Australia, 2013.

Macmahon, M. K. C. "The International Phonetic Association: The First 100 Years." *Journal of the International Phonetic Association* 16, no. 1 (June 1986): 30–38.

Madison, Paige. "All Things Bleak and Bare beneath a Brazen Sky: Practice and Place in the Analysis of Australopithecus." *History and Philosophy of the Life Sciences* 41, no. 2 (April 23, 2019): 19. https://doi.org/10.1007/s40656-019-0258-x.

Madley, Benjamin. "From Africa to Auschwitz: How German South West Africa Incubated Ideas and Methods Adopted and Developed by Nazis in Eastern Europe." *European History Quarterly* 35, no. 3 (July 2005): 429–464.

Mageo, Jeannette, and Bruce Knauft, eds. *Authenticity and Authorship in Pacific Island Encounters: New Lives of Old Imaginaries.* Berghahn Books, 2021.

Maitzen, Rohan Amanda. *Gender, Genre, and Victorian Historical Writing.* Taylor and Francis, 1998.

Malthus, T. R. *An Essay on the Principle of Population; or, a View of Its Past and Present Effects on Human Happiness; with an Inquiry into Our Prospects Respecting the Future Removal or Mitigation of the Evils Which It Occasions.* T. Bensley, 1807.

Mandler, Peter. *Return from the Natives.* Yale University Press, 2013.

Manganaro, Christine Leah. "Assimilating Hawai'i: Racial Science in a Colonial 'Laboratory,' 1919–1939." PhD diss., University of Minnesota, 2012. http://conservancy.umn.edu/handle/11299/134014.

Marks, Jonathan. "The Origins of Anthropological Genetics." *Current Anthropology* 53, no. S5 (April 2012): S161–S172. https://doi.org/10.1086/662333.

Marks, Kathy. *Lost Paradise: From Mutiny on the Bounty to a Modern-Day Legacy of Sexual Mayhem, the Dark Secrets of Pitcairn Island Revealed.* Simon and Schuster, 2009.

Maude, H. C., and H. E. Maude. "Adoption in the Gilbert Islands." *Journal of the Polynesian Society* 40, no. 4(160) (1931): 225–235.

Maude, H. C., and H. E. Maude. "The Social Organization of Banaba or Ocean Island, Central Pacific." *Journal of the Polynesian Society* 41, no. 4(164) (1932): 262–301.

Maude, H. C., and H. E. Maude. "String-Figures from the Gilbert Islands." *Journal of the Polynesian Society* 45, no. 2(178) (1936): 1–16.

Maude, Henry Evans. *Colonization of the Phoenix Islands by the Surplus Population of the Gilbert and Ellice Islands.* Government Press, 1937.

Maude, Henry Evans. "George Hunn Nobbs (1799–1884)." In *Australian Dictionary of Biography,* vol. 2, 288–289. Melbourne University Press, 1967.

Maude, Henry Evans. "The History of Pitcairn Island." In Alan S. C. Ross and A. W. Moverley, *The Pitcairnese Language,* 45–101. Andre Deutsch, 1964.

Maude, Henry Evans. "In Search of a Home: From the Mutiny to Pitcairn Island (1789–1790)." *Journal of the Polynesian Society* 67, no. 2 (June 1958): 104–131.

Maude, Henry Evans. *Of Islands and Men.* Oxford University Press, 1968.

Maude, Henry Evans. "Pacific History—Past, Present and Future." *Journal of Pacific History* 6, no. 1 (January 1971): 3–24. https://doi.org/10.1080/00223347108572180.

Maude, Henry Evans. "The Pitcairn Commemorative Stamp Issue, 1961." *Crown Agents Stamp Bulletin,* no. 346 (1960): 20–24.

Maude, Henry Evans. "Pitcairn Island." In *Encyclopedia Britannica,* vol. 17, 1116. Encyclopedia Britannica, 1967.

Maude, Henry Evans. "Review of Adams of the Bounty by Erle Wilson." *Journal of the Polynesian Society* 69, no. 2 (1960): 177–178.

Maude, Henry Evans. "Review of Pandora's Last Voyage by Geoffrey Rawson." *Journal of the Polynesian Society* 73, no. 2 (1964): 240–241.

Maude, Henry Evans. "Review of Who Caused the Mutiny on the Bounty? By Madge Darby." *Journal of Pacific History* 1 (1966): 246–247.

Maude, Henry Evans. "Some Quotations about the Pitcairnese Language." In Alan S. C. Ross and A. W. Moverley, *The Pitcairnese Language,* 118–120. Andre Deutsch, 1964.

Maude, Henry Evans. "Tahitian Interlude: The Migration of the Pitcairn Islanders to the Motherland in 1831." *Journal of the Polynesian Society* 68, no. 2 (1959): 115–140.

Maude, Henry Evans. "The Voyage of the Pandora's Tender." *Mariner's Mirror* 50, no. 3 (1964): 217–235.

Mayhew, Robert J. *Malthus: The Life and Legacies of an Untimely Prophet.* Harvard University Press, 2014.

McAdam, Jane. "Historical Cross-Border Relocations in the Pacific: Lessons for Planned Relocations in the Context of Climate Change." *Journal of Pacific History* 49, no. 3 (July 3, 2014): 301–327. https://doi.org/10.1080/00223344.2014.953317.

McAdam, Jane. "Self-Determination and Self-Governance for Communities Relocated across International Borders: The Quest for Banaban Independence." *International Journal on Minority and Group Rights* 24, no. 4 (November 3, 2017): 428–466. https://doi.org/10.1163/15718115-02404001.

McClintock, Anne. *Imperial Leather: Race, Gender, and Sexuality in the Colonial Contest.* Routledge, 1995.

McEvoy, B. P., Z. Z. Zhao, S. Macgregor, C. Bellis, R. A. Lea, H. Cox, G. W. Montgomery, L. R. Griffiths, and P. M. Visscher. "European and Polynesian Admixture in the Norfolk Island Population." *Heredity* 105, no. 2 (2010): 229–234. https://doi.org/10.1038/hdy.2009.175.

McFarland, Alfred. *Mutiny in the "Bounty" and Story of the Pitcairn Islanders.* J. J. Moore, 1884.

McGregor, Russell. "The Doomed Race: A Scientific Axiom of the Late Nineteenth Century." *Australian Journal of Politics and History* 39, no. 1 (April 1, 1993): 14–22.

Mead, Margaret. *Coming of Age in Samoa: A Psychological Study of Primitive Youth for Western Civilization.* Morrow, 1928.

"Memorabilia of the Exhibition Season." *Fraser's Magazine* 40, no. 260 (August 1851): 1–2.

Milestone, Lewis, dir. *Mutiny on the Bounty.* MGM, 1962.

Mikels-Carrasco, Jessica. "Sherwood Washburn's New Physical Anthropology: Rejecting the 'Religion of Taxonomy.'" *History and Philosophy of the Life Sciences* 34, no. 1–2 (2012): 79–101.

Milton, Hook. "Gates, Edward Harmon (1855–1940)." In *Encyclopedia of Seventh-Day Adventists.* Published September 5, 2020. https://encyclopedia.adventist.org/article?id=A9C7.

Mishra, Margaret. "'Your Woman Is a Very Bad Woman': Revisiting Female Deviance in Colonial Fiji." *Journal of International Women's Studies* 17, no. 4 (July 31, 2016): 67–78.

Mitford, Mary Russell. *Christina, the Maid of the South Seas; a Poem.* A. J. Valpy, 1811.

Mitford, Mary Russell. *The Life of Mary Russell Mitford: Told by Herself in Letters to Her Friends.* Harper and Brothers, 1870.

Mitford, Mary Russell. *Our Village: Sketches of Rural Character and Scenery.* E. Bliss, 1828.

Mitford, Nancy, ed. *Noblesse Oblige: An Enquiry into the Identifiable Characteristics of the English Aristocracy.* Hamish Hamilton, 1956.

Montagu, Ashley. "The Creative Power of Ethnic Mixture." *Psychiatry* 5, no. 4 (1942): 523–536.

Montagu, Ashley. *Man's Most Dangerous Myth: The Fallacy of Race.* Columbia University Press, 1945.

Moore, Sally Falk. "Part of the Story: A Memoir." *Ethnos* 70, no. 4 (December 1, 2005): 538–566.

Morris, Rosalind C. *Can the Subaltern Speak? Reflections on the History of an Idea.* Columbia University Press, 2010.

Morris-Reich, Amos. "Anthropology, Standardization and Measurement: Rudolf Martin and Anthropometric Photography." *British Journal for the History of Science* 46, no. 3 (September 2013): 487–516.

Moverley, Maida. *Our Pitcairn Story.* Published by the author, 2007.

Mühlhäusler, Peter. "The College of St. Barnabas on Norfolk Island and Its Languages: An Early Example of Missionary Language Planning." *Language and Communication* 30, no. 4 (October 2010): 225–239. https://doi.org/10.1016/j.langcom.2010.02.001.

Mühlhäusler, Peter. "From Despised Jargon to Language of Education: Recent Developments in the Teaching of Norf'k." In *Education in Languages of Lesser Power: Asian-Pacific Perspectives,* edited by Craig Alan Volker and Fred E. Anderson, 223–241. John Benjamins, 2015.

Mühlhäusler, Peter. "A Language Plan for Norfolk Island." In *Language Endangerment and Language Maintenance,* edited by David Bradley and Maya Bradley. Routledge Curzon, 2002.

Mühlhäusler, Peter. *Pitkern-Norf'k.* Dialects of English 17. De Gruyter Mouton, 2020.

Mühlhäusler, Peter. "Some Notes on the Ontology of Norf'k." *Language Sciences* 33, no. 4 (July 2011): 673–679.

Mühlhäusler, Peter, Rachel Nebauer-Borg, and Piria Coleman. *Ucklun's Norf'k: Words as a Memory of Our Past.* Norfolk Island Museum Trust, 2012.

Munro, Doug. *The Ivory Tower and Beyond: Participant Historians of the Pacific.* Cambridge Scholars Publishing, 2009.

Munro, Doug, and Brij V. Lal. *Texts and Contexts: Reflections in Pacific Islands Historiography.* University of Hawai'i Press, 2005.

Munslow, Alun. *A History of History.* Routledge, 2012.

Murray, Thomas Boyles. *The Home of the Mutineers.* American Sunday-School Union, 1854.

Murray, Thomas Boyles. *Pitcairn, the Island, the People, and the Pastor: To Which Is Added a Short Notice of the Original Settlement and Present Condition of Norfolk Island.* Society for Promoting Christian Knowledge, 1857.

"The Mutineers of the Bounty." *Literary World* 1, no. 2 (April 1, 1871).

Nash, Joshua. "Inside(r)-Outside(r): Toward a Linguistics and Sociology of Space on Pitcairn Island." *Australian Folklore* 31 (December 2, 2016). http://journals.kvasirpublishing.com/af/article/view/372.

Nash, Joshua. *Insular Toponymies: Pristine Place-Naming on Norfolk Island, South Pacific and Dudley Peninsula, Kangaroo Island.* John Benjamins, 2013.

Nash, Joshua. "Island Placenaming and Insular Toponymies." *Names* 63, no. 3 (September 1, 2015): 146–157. https://doi.org/10.1179/0027773815Z.000000000110.

Nash, Joshua, and Mitchell Low. "Language and Place-Knowledge on Norfolk Island." *Ethnos* 80, no. 3 (2015): 385–408.

Nechtman, Tillman W. *The Pretender of Pitcairn Island: Joshua W. Hill—The Man Who Would Be King among the Bounty Mutineers.* Cambridge University Press, 2018.

Nordhoff, Charles, and James Norman Hall. *Men against the Sea.* Little, Brown, 1934.

Nordhoff, Charles, and James Norman Hall. *Mutiny on the Bounty.* Little, Brown, 1932.

Nordhoff, Charles, and James Norman Hall. *Pitcairn's Island.* Little, Brown, 1934.

Nott, Josiah. *Two Lectures on the Natural History of the Caucasian and Negro Races.* Dade and Thompson, 1844.

Nunn, Patrick D. "Environmental Catastrophe in the Pacific Islands around AD 1300." *Geoarchaeology* 15, no. 7 (2000): 715–740.

O'Collins, Maev. *An Uneasy Relationship: Norfolk Island and the Commonwealth of Australia.* Australian National University Press, 2010.

Oliver, Dawn, ed. *Justice, Legality and the Rule of Law: Lessons from the Pitcairn Prosecutions.* Oxford University Press, 2009.

Palladino, Mariangela, and John Miller, eds. *The Globalization of Space.* Pickering and Chato, 2015.

Palliser, H. Bury. "Letter to the Secretary of the Admiralty, April 25, 1898, Enclosure No. 1." In *Pitcairn Island: Correspondence Relating to the Condition of the Pitcairn Islanders.* C. 9148. HMSO, 1899.

Palmer, Beryl Nobbs. *A Dictionary of Norfolk Words and Uses.* Published by the author, 1986.

"Papers Related to Her Majesty's Colonial Possessions, Reports for 1883, 1884, and 1885." *British Parliamentary Papers.* HMSO, 1886.

Parnell, J. W. *Norfolk Island: Report of the Administrator.* Government of Australia, 1923.

Peard, George. *The Journal of Lieutenant George Peard of H.M.S. "Blossom," 1825–1828.* Edited by Barry M. Gough. Cambridge University Press, 1973.

Pelto, Gretal, and Pertti Pelto. *The Human Adventure: An Introduction to Anthropology.* Macmillan, 1976.

Pick, Daniel. *Faces of Degeneration: A European Disorder, 1848–1918.* Cambridge University Press, 1989.

Pilling, H. G. *Report on a Visit to Pitcairn Island.* HMSO, 1930.

"Pioneer of Norfolk Island." *The Mercury,* March 19, 1934.

Pipon, Philip. *The Descendants of the Bounty's Crew.* Henry Colburn, 1834.

"Pitcairn Island." *Penny Illustrated Paper and Illustrated Times,* no. 1034 (May 7, 1881).

"Pitcairn Refutes Inbreeding Harm: Biologist's Study of Island Records Shows Mental and Physical Vigor Unimpaired." *New York Times,* November 7, 1936.

"A Pitcairner in London." *Launceston Examiner,* July 7, 1881.

"Pitcairn's Island—The Bounty's Crew." In *Colburn's United Service Magazine and Naval and Military Journal,* vol. 2, 589–593. Henry Colburn and Richard Bentley, 1829.

Power, Helen. "Pitcairn Island: Sexual Offending, Cultural Difference, and Ignorance of the Law." *Criminal Law Review* (July 2007): 609.

Pratt, Mary Louise. *Imperial Eyes: Travel Writing and Transculturation.* Routledge, 1992.

Presser, Brandon. *The Far Land: 200 Years of Murder, Mania, and Mutiny in the South Pacific.* PublicAffairs, 2022.

Prochnau, William, and Laura Parker. "Trouble in Paradise." *Vanity Fair,* December 17, 2007. https://www.vanityfair.com/news/2008/01/pitcairn200801.

Proctor, Robert, and Londa L. Schiebinger, eds. *Agnotology: The Making and Unmaking of Ignorance.* Stanford University Press, 2008.

Psomiades, Kathy Alexis. "Polygenist Ecosystems: Robert Knox's *The Races of Man* (1850)." *Victorian Review* 36, no. 2 (2010): 32–36.

Qureshi, Sadiah. *Peoples on Parade: Exhibitions, Empire, and Anthropology in Nineteenth-Century Britain.* University of Chicago Press, 2011.

Rabinow, Paul. *Reflections on Fieldwork in Morocco.* University of California Press, 1977.

The Race Concept: Results of an Inquiry. UNESCO, 1952.

"A Racial Laboratory." *New York Times,* February 28, 1935.

Radin, Joanna. *Life on Ice: A History of New Uses for Cold Blood.* University of Chicago Press, 2017.

Raj, Kapil. *Relocating Modern Science: Circulation and the Construction of Knowledge in South Asia and Europe, 1650–1900.* Palgrave Macmillan, 2007.

Rapaport, Moshe. *The Pacific Islands: Environment and Society.* Rev. ed. University of Hawai'i Press, 2013.

Reardon, Jenny, and Kim TallBear. "'Your DNA Is Our History': Genomics, Anthropology, and the Construction of Whiteness as Property." *Current Anthropology* 53, no. S5 (April 1, 2012): S233–S245. https://doi.org/10.1086/662629.

Reinecke, John E., ed. *A Bibliography of Pidgin and Creole Languages.* University of Hawai'i Press, 1975.

Reinecke, John E. "Marginal Languages." PhD diss., Yale University, 1937.

Reinecke, John E. "Pitcairnese and Norfolk." In *A Bibliography of Pidgin and Creole Languages,* edited by John. E. Reinecke, 590–592. University of Hawai'i Press, 1975.

"Relics of the Past in London." *Nation* 60, no. 1558 (May 9, 1895): 357–358.

Rennie, Neil. *Far-Fetched Facts: The Literature of Travel and the Idea of the South Seas.* Clarendon, 1995.

Review of *The True Law of Population,* by Thomas Doubleday. *British Quarterly Review* 2, no. 4 (August 1845): 442.

Reynolds, Pauline. "The Forgotten Women of the 'Bounty' and Their Material Heritage." In *Conference Proceedings: Maori and Pacific Textile Symposium: Whatu Raranga a Kiwa, Understanding and Uniting Mori and Pacific Textiles.* Museum of New Zealand Te Papa Tongarewa, 2012.

Reynolds, Pauline. "Pitcairn Tapa: Discovering the Past through Tapa." In *TAPA: From Tree Bark to Cloth: An Ancient Art of Oceania from Southeast Asia to Eastern Polynesia*, edited by Michel Charleux, 370–375. Somogy Art Publishers, 2017.

Reynolds, Pauline. "Tapa Cloths and Beaters: Tradition, Innovation and the Agency of the Bounty Women in Shaping a New Culture on Pitcairn Island from 1790 to 1850." *Textile History* 47, no. 2 (July 2, 2016): 190–207. https://doi.org/10.1080/00404969.2016.1211435.

Reynolds, Pauline, and Jean Clarkson. "'Ahu Sistas: Reclaiming History, Telling Our Stories." In *Material Approaches to Polynesian Barkcloth: Cloth, Collections, Communities*, ed. Frances Lennard and Andy Mills, 191–199. Sidestone Press, 2020.

Rodwell, Cecil Hunter. *Report on a Visit to Pitcairn Island.* HMSO, 1921.

Rogner, Emil. *Pitcairn-Island: Ein Beitrag zur Lehre von der Entfaltung menschlicher Gesellungen* [Pitcairn Island: A contribution to the study of the development of human society]. Nürnberger Beiträge zu den Wirtschafts- und Sozialwissenschaften [Nuremberg contributions to economics and social science] 78. Verlag der Hochschulbuchhandlung, Krisdie, 1940.

"The Romance of the Sea Ended." *The Evangelist,* March 9, 1899.

Ross, Alan S. C. "Linguistic Class-Indicators in Present-Day English." *Neuphilologische Mitteilungen* 55, no. 1 (1954): 20–56.

Ross, Alan S. C., and A. W. Moverley. "Note on Some Pristine Place-Names of Pitcairn Island." In *Cinquième congrès international de toponymie et d'anthroponymie, Salamanca, 12–15 avril 1955: Actes et mémoires* [Fifth international congress of toponymy and anthroponymy, Salamanca, April 12–15, 1955: Proceedings and transactions]. Vol. 2, edited by L. Cortés, M. García Blanco, and A. Tovar. University of Salamanca, 1955.

Ross, Alan S. C., and A. W. Moverley. *The Pitcairnese Language.* Andre Deutsch, 1964.

"Royal Geographical Society." *Athenaeum* 505 (July 1837): 484.

Rozett, Tamar I. "The Case of Claud Cardew's Violin: Race, Anxiety, and the British Empire Mail." *Journal of British Studies* 61, no. 3 (July 2022): 599–620. https://doi.org/10.1017/jbr.2022.53.

Safier, Neil. "Global Knowledge on the Move: Itineraries, Amerindian Narratives, and Deep Histories of Science." *Isis* 101, no. 1 (March 2010): 133–145. https://doi.org/10.1086/652693.

Salesa, Damon Ieremia. *Racial Crossings: Race, Intermarriage, and the Victorian British Empire.* Oxford University Press, 2011.

Sanders, Roy. "Our Island, Being a Study of Pitcairn Community in Transition." Master's thesis, University of Auckland, 1953.

Sargent, Charles Lenox, and John Adams. *The Life of Alexander Smith, Captain of the Island of Pitcairn, One of the Mutineers on Board His Majesty's Ship Bounty, Commanded by Lieut. Wm. Bligh.* Printed by S. T. Goss, 1819.

Satia, Priya. *Time's Monster.* Harvard University Press, 2020.

Saxer, Martin, and Andersson, Ruben. "The Return of Remoteness: Insecurity, Isolation and Connectivity in the New World Disorder." *Social Anthropology* 27, no. 2 (2019): 140–155. https://doi.org/10.1111/1469-8676.12652.

Schneider, Axel, and Daniel Woolf. *The Oxford History of Historical Writing.* Vol. 5, *Historical Writing since 1945.* University Press, 2011.

Schubert, Ernest. *The Pitcairn Island Story.* Longmans, 1961.

Schubert, Ernest, Alan S. C. Ross, and A. W. Moverley. "Pitcairn Island Today." In Alan S. C. Ross and A. W. Moverley, *The Pitcairnese Language,* 26–36. Andre Deutsch, 1964.

"Scientific Survey of South-Eastern Polynesia." *Nature* 133, no. 3362 (April 7, 1934): 527.

Scott, James C. *Domination and the Arts of Resistance: Hidden Transcripts.* Yale University Press, 2008.

Secord, James A. "Knowledge in Transit." *Isis* 95, no. 4 (December 1, 2004): 654–672. https://doi.org/10.1086/430657.

Sekora, John. "Black Message/White Envelope: Genre, Authenticity, and Authority in the Antebellum Slave Narrative." *Callaloo,* no. 32 (1987): 482–515. https://doi.org/10.2307/2930465.

Seth, Suman. *Difference and Disease: Medicine, Race, and the Eighteenth-Century British Empire.* Cambridge University Press, 2018.

Shapiro, Harry L. "Forensic Anthropology." *Annals of the New York Academy of Sciences* 318, no. 1 (February 2, 1979): 3–9.

Shapiro, Harry L. *Descendants of the Mutineers of the Bounty.* Vol. 11, no. 1 of *Memoirs of the Bernice P. Bishop Museum*. Bishop Museum Press, 1929.

Shapiro, Harry L. *The Heritage of the Bounty: The Story of Pitcairn through Six Generations.* Simon and Schuster, 1936.

Shapiro, Harry L. *The Heritage of the Bounty.* Rev. with new postscript. Anchor Books, 1962.

Shapiro, Harry L. *The Heritage of the Bounty.* AMS Press, 1979.

Shapiro, Harry L. *The Pitcairn Islanders (Formerly the Heritage of the Bounty).* Simon and Schuster, 1968.

Shapiro, Harry L. *The Pitcairn Islanders.* Rev. ed. Simon and Schuster, 1986.

Shapiro, Harry L. "Race Mixture." In *The Race Question in Modern Science.* Columbia University Press, 1961.

Shapiro, Harry L. "Results of Inbreeding on Norfolk Island." *Science* 65, no. 1693 (1927): 10.

Shapiro, Harry L. "Revised Version of UNESCO Statement on Race." *American Journal of Physical Anthropology* 10, no. 3 (September 3, 1952): 335–336.

Shapiro, Harry L. "The Romance of the Norfolk Islanders: Modern Descendants of the Mutineers Furnish Material for Research." *Scientific American* 135, no. 3 (September 1926): 182–184.

Shapiro, Harry L. "Some Reflections of an Anthropologist on the Future of Our Population." *Proceedings of the American Philosophical Society* 80, no. 4 (February 15, 1939): 587–600.

Shapiro, Harry L. "A Study of Race Mixture as Exemplified in the Descendants of Tahitian and English Mutineers of the Bounty." PhD diss., Harvard University, 1926.

Sherwin, Justin C., Lisa S. Kearns, Alex W. Hewitt, Yaling Ma, John Kelly, Lyn R. Griffiths, and David A. Mackey. "Prevalence of Chronic Ocular Diseases in a Genetic Isolate: The Norfolk Island Eye Study (NIES)." *Ophthalmic Epidemiology* 18, no. 2 (2011): 61–71. https://www.tandfonline.com/toc/iope20/18/2.

Shillibeer, John. *A Narrative of the Briton's Voyage, to Pitcairn's Island.* Law and Whittaker, 1817.

The Ship and the Island. Groombridge and Sons, 1849.

Shipman, Pat. *The Evolution of Racism: Human Differences and the Use and Abuse of Science.* Harvard University Press, 1994.

Sinoto, Yosihiko H. "An Analysis of Polynesian Migrations Based on the Archaeological Assessments." *Océanistes* 39 (1983): 57–67.

Sivasundaram, Sujit. *Islanded: Britain, Sri Lanka, and the Bounds of an Indian Ocean Colony.* University of Chicago Press, 2013.

Smith, Bonnie G. *The Gender of History: Men, Women, and Historical Practice.* Harvard University Press, 2000.

Smith, Vanessa. *Intimate Strangers: Friendship, Exchange and Pacific Encounters.* Cambridge University Press, 2010.

Smith, Vanessa. "Pitcairn's Guilty Stock: The Island as Breeding Ground." In *Islands in History and Representation,* edited by Vanessa Smith and Rod Edmond, 116–132. Routledge, 2003.

Spencer, Frank. *Ecce Homo: An Annotated Bibliographic History of Physical Anthropology.* Greenwood Press, 1986.

Spivak, Gayatri Chakravorty. "Can the Subaltern Speak?" In *Marxism and the Interpretation of Culture,* edited by Cary Nelson and Lawrence Grossberg, 271–313. Macmillan, 1988.

"Sport and Natural History in Norfolk Island." *Country Life,* April 11, 1903.

Spriggs, Matthew. "Are Islands Islands? Some Thoughts on the History of Chalk and Cheese." In *Islands of Inquiry: Colonisation, Seafaring and the Archaeology of Maritime Landscapes,* edited by Geoffrey Clark, Foss Leach, and Sue O'Connor. Terra Australis 29. Australian National University Press, 2008. https://www.jstor.org/stable/j.ctt24h8gp.16.

Stanley, E. G. "Ross, Alan Strode Campbell (1907–1980)." In *Oxford Dictionary of National Biography,* edited by H. C. G. Matthew and Brian Harrison. Oxford University Press, May 2, 2004.

Steley, Dennis. "Unfinished: The Seventh-Day Adventist Mission in the South Pacific, Excluding Papua New Guinea, 1886–1986." PhD diss., University of Auckland, 1989.

Stepan, Nancy. *The Idea of Race in Science: Great Britain, 1800–1960.* Macmillan, 1982.

Stevens, Kate. *Gender, Violence and Criminal Justice in the Colonial Pacific: 1880–1920.* Bloomsbury Publishing, 2022.

Stewart, Charles, ed. *Creolization: History, Ethnography, Theory.* Left Coast Press, 2007.

Stocking, George W. *Observers Observed: Essays on Ethnographic Fieldwork.* University of Wisconsin Press, 1984.

Stoler, Ann Laura. *Along the Archival Grain: Epistemic Anxieties and Colonial Common Sense.* Princeton University Press, 2010.

Stoler, Ann Laura. *Carnal Knowledge and Imperial Power: Race and the Intimate in Colonial Rule.* University of California Press, 2002.

Storey, Kenton. *Settler Anxiety at the Outposts of Empire: Colonial Relations, Humanitarian Discourses, and the Imperial Press.* University of British Columbia Press, 2016.

Tainter, Joseph A. "The Archaeology of Overshoot and Collapse." *Annual Review of Anthropology* 35 (2006): 59–74.

Teaiwa, Katerina Martina. *Consuming Ocean Island: Stories of People and Phosphate from Banaba.* Indiana University Press, 2014.

Teaiwa, Katerina Martina. "Our Sea of Phosphate: The Diaspora of Ocean Island." In *Indigenous Diasporas and Dislocations,* edited by Charles D. Thompson and Graham Harvey, 169–192. Routledge, 2017.

Teaiwa, Katerina Martina. "Ruining Pacific Islands: Australia's Phosphate Imperialism." *Australian Historical Studies* 46, no. 3 (September 2, 2015): 374–391. https://doi.org/10.1080/1031461X.2015.1082609.

Teaiwa, Teresia K. "L(o)osing the Edge." *Contemporary Pacific* 13, no. 2 (2001): 343–357.

Teehuteatuaonoa (Jenny). "Account of Jenny, an Otahetian Woman." *United Services Journal* 2, no. 2 (November 1829): 589–593.

Teehuteatuaonoa (Jenny). "Account of the Mutineers of the Ship Bounty, and Their Descendants at Pitcairn's Island." *Sydney Gazette and New South Wales Advertiser*, July 17, 1819. http://nla.gov.au/nla.news-article2178824.

Terrell, John Edward. "Metaphor and Theory in Island Archaeology." *Journal of Island and Coastal Archaeology* 8, no. 4 (October 29, 2020): 1–11. https://doi.org/10.1080/15564894.2020.1830892.

Teslow, Tracy. *Constructing Race: The Science of Bodies and Cultures in American Anthropology*. Cambridge University Press, 2014.

Thomas, Nicholas. *Entangled Objects: Exchange, Material Culture, and Colonialism in the Pacific*. Harvard University Press, 1991.

Thomas, Nicholas. *Out of Time: History and Evolution in Anthropological Discourse*. University of Michigan Press, 1989.

"Three Months on Pitcairn." *Journey of a Lifetime*. BBC Radio 4, October 19, 2015. https://www.bbc.co.uk/programmes/b06j144b.

Tilley, Helen. *Africa as a Living Laboratory: Empire, Development, and the Problem of Scientific Knowledge, 1870–1950*. University of Chicago Press, 2011.

Torstendahl, Rolf. *The Rise and Propagation of Historical Professionalism*. Routledge, 2014.

The Transformed Island: A Story of the South Seas. Presbyterian Board of Publication, 1854.

Trenwith, Anthony. "The Empire Strikes Back: Human Rights and the Pitcairn Proceedings." *Journal of South Pacific Law* 7, no. 2 (2003).

Trevor, J. C. "Some Anthropological Characteristics of Hybrid Populations." *Eugenics Review* 30, no. 1 (April 1930): 21–31.

Trevor-Roper, Hugh. *The Rise of Christian Europe*. Thames and Hudson, 1965.

Tsing, Anna Lowenhaupt. "From the Margins." *Cultural Anthropology* 9, no. 3 (August 1994): 279–297.

Tuck, Eve, and K. Wang Yang. "R-Words: Refusing Research." In *Humanizing Research: Decolonizing Qualitative Inquiry with Youth and Communities*, edited by Django Paris and Maisha T. Winn. Sage, 2013.

Turner, G. W. Review of *The Pitcairnese Language*, by A. S. C. Ross and A. W. Moverley. *Anglia* 86 (1968): 358–360.

"Unity of the Races." *Southern Quarterly Review* 7, no. 14 (April 14, 1845): 446.

Vetter, Jeremy. "Wallace's Other Line: Human Biogeography and Field Practice in the Eastern Colonial Tropics." *Journal of the History of Biology* 39, no. 1 (March 2006): 89–123.

"Voyage de Dentrecasteaux." *Quarterly Review* 3, no. 5 (February 1810): 21–43.

Wagner, Kim A. "'Treading upon Fires': The 'Mutiny'-Motif and Colonial Anxieties in British India." *Past and Present* 218, no. 1 (February 1, 2013): 159–197. https://doi.org/10.1093/pastj/gts026.

Wagner, K. "The Variability of Hybrid Populations." *American Journal of Physical Anthropology* 16, no. 3 (1932): 283–307.

Wallace, Alfred Russel. "The Mutineers of the Bounty, and Their Descendants in Pitcairn and Norfolk Islands, by Lady Belcher." *The Academy* (February 1, 1871): 107–109. http://people.wku.edu/charles.smith/wallace/S182A.htm.

Washburn, Sherwood. "Section of Anthropology: The New Physical Anthropology." *Transactions of the New York Academy of Sciences* 13, no. 7 (May 1951): 298–304.

Weisler, Marshall I. "Henderson Island Prehistory: Colonization and Extinction on a Remote Polynesian Island." *Biological Journal of the Linnean Society* 56, no. 1–2 (1995): 377–404. https://doi.org/10.1111/j.1095-8312.1995.tb01099.x.

White, Hayden. *Metahistory: The Historical Imagination in Nineteenth-Century Europe.* 1974; repr. Johns Hopkins University Press, 2014.

Wiercinski, Andrzej, and Tadeusz Bielicki. "The Racial Analysis of Human Populations in Relation to Their Ethnogenesis." *Current Anthropology* 3, no. 1 (February 1962): 2, 9–46.

Wilson, Kathleen. *The Island Race: Englishness, Empire and Gender in the Eighteenth Century.* Routledge, 2003.

Woodburn, Susan. *Where Our Hearts Still Lie: Harry and Honor Maude in the Pacific Islands.* Crawford House Publishing Australia, 2003.

"A Word for the Reading Course." *North Pacific Union Gleaner,* September 21, 1910.

Wurm, S. A., and D. C. Laycock. "The Question of Language and Dialect in New Guinea." *Oceania* 32, no. 2 (December 1961): 128–143.

Yarbough, Jean, dir. *The Women of Pitcairn Island.* 20th Century Fox, 1956.

Young, Adrian. "Bounty Relics: Trading in the Legacy of Myth and Mutiny." In *The Bounty from the Beach: Cross-Cultural and Cross-Disciplinary Essays,* edited by Sylvie Largeaud-Ortega, 207–34. Australian National University Press, 2018.

Young, Adrian. "Putting Anthropology in Its (Hospitable) Place: Harry Shapiro's Fieldwork on Pitcairn Island, 1934–1935." *History and Anthropology* 33, no. 5 (October 20, 2022): 631–651. https://doi.org/10.1080/02757206.2020.1762591.

Young, Adrian, Maria Amoamo, Martin Gibbs, Alexander Mawyer, Joshua Nash, Tillman W. Nechtman, and Pauline Reynolds. "When Margins Are Centres: De-Ranging Pitcairn Island's Place in Pacific Scholarship." *Journal of the Polynesian Society* 130, no. 3 (September 30, 2021): 197–226.

Young, Robert. *Colonial Desire: Hybridity in Theory, Culture, and Race.* Routledge, 1995.

Young, Rosalind Amelia. "How the Pitcairn Islanders Spend Their Sunday." *The Advance,* September 14, 1882.

Young, Rosalind Amelia. "Letters from Pitcairn Island." *Overland Monthly and Out West Magazine* 18, no. 105 (September 1891): 294–308.

Young, Rosalind Amelia. *Mutiny of the Bounty and Story of Pitcairn Island, 1790–1894.* Pacific Press, 1894.

Young, Rosalind Amelia. "News from Pitcairn." *Signs of the Times,* November 19, 1906.

Young, Rosalind Amelia. "Pitcairn Island, Aug. 26, 1881." *Century Illustrated Monthly Magazine* 24, no. 2 (1882).

Zettersten, Arne. "The Mutineers of the *Bounty* and the Language of Their Descendants." *Moderna Sprak* 60 (1966): 393–398.

Index

About the Author

Adrian Young is a historian of Europe and the world and an associate professor at Denison University, with a long-standing interest in the history of knowledge making and colonialism. He earned a PhD in the history of science at Princeton University, and has published articles and book chapters on the history of anthropology, material culture, and the supposed marginality of Pacific islands in journals such as *History and Anthropology* and *Waka Kuaka: The Journal of the Polynesian Society.*